Dunbar Cave

The Showplace of the South

Dunbar Cave

The Showplace of the South

Larry E. Matthews

Library of Congress Control Number 2011932077
ISBN 978-1-879961-41-8

Second Edition July, 2011

Produced by the
National Speleological Society
Special Publications Committee
G. Thomas Rea, Chairman

Front Cover photograph: Larry E. Matthews and Amy Wallace relax in front of the Entrance to Dunbar Cave. Photo by Bob Biddix, May 27, 2011.

Back Cover illustration: 1905 colorized postcard of the Entrance to Dunbar Cave from the Billyfrank Morrison Collection. Note the wooden dance floor.

Book Design by
Greyhound Press
Bloomington, Indiana

GREYHOUND PRESS

Published by
The National Speleological Society
2413 Cave Avenue
Huntsville, Alabama 35810-4431
USA

For my Mother

Dorothy Tyler Matthews

Born: December 28, 1917

Died: March 23, 2009

My mother took the time to take my friends and me to places where we could collect minerals and fossils while we were still in Junior High School. Then, when we discovered there were caves out there, she took us to dozens of wild caves before we were old enough to have our own driver's licenses. She spent many a day sitting in her car, while we were exploring a cave.

My mother was born in Clarksville, Tennessee and, therefore, was very familiar with Dunbar Cave's social history. While conducting research in the cave, I found names on the cave walls that had the last name Tyler, her maiden name. I took photos of these names and showed them to her, in hopes that some of them might be long-lost relatives. But, she was unable to recognize any of the names. Her grandfather was Judge Charles Waller Tyler, the County Judge from August 4, 1873 to September 1, 1918. The area of Clarksville where he lived is still known as Tylertown.

Several times we went swimming at Dunbar Cave while I was a small child. I was hoping to find a photograph of one of those family outings and include it in this Second Edition of the book, but all my family photographs were destroyed in the Great Flood here in Nashville, Tennessee, on May 2, 2010.

Table of Contents

Dunbar Cave

Acknowledgements for the First Edition

First and foremost, the history of the modern exploration of Dunbar Cave has been intimately tied to the history of the Northern Indiana Grotto of the National Speleological Society. This intrepid group of explorers not only made many exciting discoveries in Dunbar Cave, but they also took the time to carefully survey and record their discoveries. Their newsletter, *The Michiana Caver*, has been an invaluable resource in writing this book. Bill Torode and the National Speleological Society Library provided me with back issues of this publication for my research.

Cavers (listed alphabetically) who have assisted in providing information and in reviewing the chapters on modern exploration are Charles Blakeway, Gary Collins, Jack Countryman, Tom Cussen, Keith Dunlap, Richard P. Geer, Dan McDowell, Karl Niles, James David Reyome, Walter G. Scheffrahn, David D. Seng, and Clyde N. Simerman. Their help has assured the accuracy of those chapters.

Dan McDowell spent countless hours proofreading the entire text, which resulted in his locating and correcting many small errors. This has greatly improved the accuracy and value of this book.

Cave biologist Thomas C. Barr Jr provided information on cave life in Dunbar Cave. Amy Atkins Wallace, the interpretive Specialist for Dunbar Cave, provided additional information on the biology of the cave.

Cave historian Marion O. Smith generously supplied me with all his research data on Dunbar Cave, including copies of his notes taken inside the cave. Cave Historian Joe Douglas pointed out numerous pieces of evidence of prehistoric exploration while we were in the cave on January 15, 2005.

Cave photographer Tom Cussen provided many of the photographs used in this book. He contributed a tremendous amount of time and effort in preparing his photographs for use and in assuring that the captions were accurate. James David Reyome and Clyde Simerman provided other photographs. These photographs add immensely to this book.

Expert caver Gerald Moni shared his observations on prehistoric use of the cave.

Geologists Elaine Foust, Robert Hall, and John Hoffelt reviewed the chapter on the geology of Dunbar Cave and made helpful suggestions. John Hoffelt also provided valuable comments on the text.

The staff of Dunbar Cave State Natural Area was very helpful while I was researching this book. These include Bob Wells, the Park Manager, Amy Atkins Wallace, the Interpretive Specialist; and Robert S. Embler, a seasonal employee. Amy Wallace also read several of the chapters and provided helpful comments. The "Dunbar Cave State Natural Area Research Project," which was prepared by Deborah Lingle Gillis, is on file at the Dunbar Cave Natural Area's Office and was a valuable source of information.

The Staff of the Tennessee Natural Areas Program provided access to their files and reviewed some of the materials in this book. These people include Brian Bowen, Forrest Evans, Reggie Reeves, and David Withers.

Charles Crow obtained permission for us to visit the site of the Idaho Springs Hotel, the Idaho Springs, and some of the original cabins.

Deborah Hammerstein helped many times with niggling little questions that needed answers.

Postcard expert and historian Billyfrank Morrison provided invaluable assistance on researching the postcards concerning Dunbar Cave and proofread Appendix B, "Postcards." The staff of the Tennessee State Library and Archives, including Susan L. Gordon and Kassandra Hassler, also helped with research on postcards as did the staff of the Nashville Room of the Public Library of Nashville and Davidson County.

Sims Crownover helped with the research on the legend concerning Jenny Lind and proofread Appendix B, "Postcards."

Louise Zepp, Editor of the *Tennessee Conservationist* magazine, provided helpful information on the copyright laws.

Amy Atkins Wallace at Dunbar Cave made numerous trips into the cave with me and provided unlimited access to the files in the Natural Area Office. Without her support, this book would not have been possible.

Tom Rea, Chairman of the National Speleological Society's Special Publications Committee, provided tremendous support by preparing my rough manuscript for publication.

Last, and not least, I would like to thank my wife Betty, for all her support over the years. Being married to a caver and an author cannot be easy.

Acknowledgements for the Second Edition

Bob Biddix provided the beautiful photo on the Front Cover. I can always count on Bob to take a photo for me when I need one.

Expert cave photographer Alan Cressler provided excellent photographs of the prehistoric Indian glyphs.

Cave historian and professional geologist Angelo I. George was kind enough to allow me to quote his observations on early saltpeter mining. He proofread the chapter on saltpeter mining and made several helpful suggestions.

Caver Carol Lavender assisted in taking photographs of other attractions in the Dunbar Cave area.

Historian Billyfrank Morrison accompanied me on numerous trips into the cave to research old names and dates and to study the Indian glyphs. He provided access to his remarkable collection of Dunbar Cave postcards (again) and allowed me to use even more of them in this Second Edition. They are a great window into the past.

My Editor, Tom Rea, is just incredible. I am always amazed at how good my work looks after he is finished with it. Little did I realize when we first worked together in 2004 and 2005 on the First Edition of this book, what a long and productive relationship this was going to be.

University of Tennessee Anthropologist Jan F. Simek offered assistance in the preparation of the material on the Indian glyphs and was kind enough to allow me to quote his material and use his photographs and drawings. He also offered helpful suggestions in editing that chapter.

Amy Atkins Wallace provided tremendous assistance and support in this project. Again, she made numerous trips into the cave with me and provided unlimited access to the files in the Natural Area Office. Without her support, this Second Edition would not have been possible.

Last, but not least, thanks again to my wife Betty for her continuing support in my research projects.

Introduction to the First Edition

Dunbar Cave--which is located inside the city limits of Clarksville, Tennessee, has been open to the public for more than 146 years. For much of that time it was a privately owned cave, but since 1973 Dunbar Cave and the surrounding 110 acres of land have been a part of the Tennessee State Park System. Dunbar Cave is known officially as a "State Natural Area." Therefore, the land above the cave is open to the general public for their enjoyment and guided tours of the cave are offered on a regular basis.

Thomas C. Barr Jr, in his famous book, *Caves of Tennessee*, described Dunbar Cave in 1961 as follows:

> Dunbar Cave has been open to the public for more than 75 years and is a noted recreational spot. The cave stream emerges below the mouth and is dammed up to form a lake. At present the cave is electrically lighted.
>
> The entrance is 35 feet wide and 10 feet high. The cave consists of three large, subparallel galleries and a number of smaller side passages. The large galleries trend north-northwest and average about 15 to 20 feet wide and 8 to 12 feet high. The main avenue of the cave is 2,000 feet or more long. About 1,500 feet from the mouth is Petersons Leap, a 35-foot pit developed by slumping of the fill into a lower level. According to tradition, a man named Peterson became lost in the cave and fell into the pit.
>
> West of the main avenue is a series of large chambers along the stream, which is wide, deep, and flows slowly, almost imperceptibly, through the cave. It is inhabited by blindfishes (*Typhlichthys subterraneus*).
>
> East of the main avenue is a series of chambers well decorated with dripstone, the most popular section of the cave. The largest of the formation rooms are Independence Hall and The Egyptian Temple. Near Petersons Leap, a crawlway extends northward for 150 feet into a seldom-visited room which contains a deep, rock-rimmed pool and a large, active flowstone formation.
>
> The main cave was explored as far as a 20-foot solution pit that extends across the floor from wall to wall. A large avenue can be seen extending beyond, but its length was not determined. The total length of passages visited by the writer is about 4,500 feet.[1]

Although the passages Barr describes are very large and interesting, the portions of the cave described by him at that time are less than 1 mile long. As it turns out, Barr did not have a chance to visit all of the known cave and a more recent map, dated 1978, that does include all the various crawlways and side passages known at that time, shows a total length of 16,145 feet, or over 3.3 miles. This portion of the cave is referred to as the Historic Section in this book.

Since 1977, a group of intrepid cave explorers has discovered miles of virgin passages in this cave and carefully mapped these discoveries to the cave's current known length of 8.08 miles. This makes Dunbar Cave the 11th

---

1   Thomas C. Barr Jr., *Caves of Tennessee*, Bulletin 64, Tennessee Division of Geology, 1961.

longest known cave in the state of Tennessee. When you consider that Tennessee has over 8,000 known caves you begin to appreciate that this is one of the premier caves in the entire state. At the present time (2004), this also makes Dunbar Cave the 79th longest mapped cave in the United States. In the world of caves, Dunbar Cave is a very large cave.

The history of Dunbar Cave, however, stretches far back into prehistoric times when American Indians frequented its entrance and left traces of their explorations far from the entrance. Recent archeological studies have shed further light on their activities. These will be discussed in Chapter 1 – American Indians.

Introduction to the Second Edition

On January 15, 2005, just a matter of weeks before the manuscript for the First Edition of *Dunbar Cave: The Showplace of the South* was sent to Tom Rea, the editor, Amy Wallace, Billyfrank Morrison, Joe Douglas, and I discovered Indian glyphs in Dunbar Cave. Clearly, this was a major discovery and would require research by trained archaeologists. None of us were archaeologists, but we did promptly report the discovery to the University of Tennessee at Knoxville. Rather than wait months, or even years, for this research to be complete, I decided to publish the First Edition with only a slight mention of this discovery on page 9. In retrospect, this was a wise decision, because over 6 years later, the research on these glyphs is still not complete and has not yet been published.

There is really no way that an exhaustive history could ever be written about a cave like Dunbar Cave. Therefore, I reached a point in writing the First Edition where I thought I conveyed the most important parts of the cave's history, especially the recent exploration of the cave by cave explorers. All in all, I was reasonably satisfied with the results when they were published in May, 2005.

However, the First Edition did not cover the social history of the cave nearly as well as it covered the history of the cave's exploration. This Second Edition is my attempt to add much more of this fascinating information to the book. Furthermore, information on the history of the cave from 1963 to 1973, when it was owned by McKay King, was almost totally missing in the First Edition. This additional social history and information on the McKay King period of ownership, along with information on the cave's very rare and interesting prehistoric art, add substantially to the knowledge of the cave. It is hoped that the reader will find the Second Edition enjoyable and informative.

Larry E. Matthews

Illustrations

Dunbar Cave

Dunbar Cave

Chapter 1

American Indians

Archeological investigations both inside and outside of Dunbar Cave indicate that the cave was used by some of the first humans to inhabit Tennessee. This is not surprising, because primitive man has always found caves and rockshelters[1] ideal sites for both temporary and permanent residences. A large cave entrance, such as the one at Dunbar Cave, provided shelter from the weather, and the spring at the entrance provided a year-round supply of fresh water. During hot weather, the air blowing out of the cave is delightfully cool, and during cold weather, the cave air is much warmer than the outside air. It could be easily argued that modern houses are artificial caves, providing similar shelter. Man had to invent houses to provide shelter where there were either not enough suitable caves to go around or to move into those areas where there were no caves. Well-known caver Gerald Moni has visited most of the more than 3.3 miles of the Historic Section of the cave that was explored before the modern era of cave exploration (which began in Dunbar Cave in the 1970s). He states that "cane stoke marks"[2] are abundant throughout

this portion of the cave, indicating that it was heavily visited by American Indians in prehistoric times. Gerald also reports finding cane torch charcoal in the cave. Cane is related to bamboo and was abundant in this area during prehistoric times. Cane torches were used only by American Indians and not by historic people, according to Moni. He estimates that American Indians saw approximately 2.5 miles of the Historic Section of the cave. The remainder was dug open after the arrival of the European settlers and was too low to explore before that time. When I asked Gerald what he thought the prehistoric people were looking for, it was his opinion that they had been recreational cavers, not miners. In some caves, such as Mammoth Cave, Kentucky, we can see where the American Indians were mining salts. But here, in Dunbar Cave, he sees no evidence of any mining activity by the Indians.

Cave historian Joe Douglas accompanied me into the cave on January 15, 2005, and pointed out numerous places where prehistoric peoples had left cane stoke marks on the walls. Chert and clay are both abundant in the cave, but no clear evidence of mining by the Indians was noted.

Several years after the state of Tennessee purchased Dunbar Cave, the state Division of Archeology evaluated it as an archeological site. In a report prepared by Victor P. Hood with the Division of Anthropology at the University of Tennessee at Knoxville, they gave the

---

1   A rockshelter is an overhanging bluff that provides protection from the weather. It usually does not extend back far enough to be out of the natural light to be considered a true cave.

2   Cane stoke marks are charcoal smudges left on the cave walls where American Indians tapped their cane torches to remove excess coals from the ends of the torch.

following assessment:

No formal archeological considerations were given to the purchase of Dunbar Cave, but for several reasons it is among the most ideal shelters for investigation in the region. Although numerous rockshelters and caves in the area exhibit evidence of extensive human occupation, the majority of these have suffered significantly in recent years from vandalism and pot hunting. Because of its popularity, Dunbar Cave was spared this all too common fate. The large entrance area of the cave, including most of the potential habitation area, was sealed in 1906 with large amounts of fill dirt and a wooden floor. Later, the addition of a concrete dance floor further insured the prevention of any

Figure 1-1. Archeological excavation, Dunbar Cave entrance, 1978. Photo courtesy of Dunbar Cave State Natural Area.

post-1906 disturbance of the cultural deposits.[3]

The Tennessee Division of Archeology conducted the first evaluation of the Dunbar Cave entrance as an archeological site on March 7–9, 1977. The personnel conducting the evaluation were Brian Butler, Robert Jolley, David High, Emanuel Breitburg, and Jim East. Six small test pits were excavated. Three of these pits were in the entrance area outside of the iron gate and three more were inside the gate. Brian M. Butler with the Division of Archeology wrote the following report of those excavations:

The excavation revealed the presence of substantial archeological deposits, particularly in the mouth of the cave. Caves are a distinctive class of archaeological site. Because of the restricted living area, refuse is not dispersed over large areas and very high concentrations of refuse and artifacts result. The densities of material are usually greater than those found on open habitation sites. Preservation of faunal remains is also generally superior to that on open sites. Both conditions hold true for Dunbar Cave. The small test holes produced large amounts of material, particularly bone and shell. The condition of most of the bone and shell is excellent. By comparison, small numbers of lithic and ceramic items were found. Diagnostic items were scarce, but this is not surprising given the small size of the test units.

Mississippian and Woodland Period occupations are indicated by the presence of pottery. One shell-tempered sherd and several limestone-tempered sherds (one cord-marked) were recovered from the upper 50 centimeters of the deposit. Several stemmed projectile points were also found; their form is vaguely suggestive of Archaic

---

3 Robert A. Pace and Victor P. Hood, "Archeological Investigtions at Dunbar Cave: A Stratified Rockshelter."

*Figure 1-2. Archaeological excavation underway in the Dunbar Cave Entrance.
Photo courtesy of Dunbar Cave State Natural Area.*

but their small size and stratigraphic position could indicate a lesser age. One expanded base Archaic drill was found. It is expected that most of the deposit will probably represent Archaic Period habitation. Caves are notorious for uneven stratigraphy and disturbances, so the provenience of the sherds in the straigraphic column need not be too significant.

Little chert debitage was recovered. Most of it consists of decertification flakes from water-rounded chert nodules. Several chert and limestone nodules exhibit use as hammerstones.

Numerically speaking, the bulk of the material recovered is bone and shell, some 500 elements. This total includes 11 pieces of bone tools or modified bone. These include several deer ulna awls, a turkey bone awl, fragments of several splinter awls, a bone "needle," an antler flaker, an antler projectile point, and residue from making a bone fish hook. This amount of worked material represents a very high concentra-

tion of such materials.

Human remains were also collected from the cave. Various portions of the disturbed burial on the interior of the cave were recovered. The bones represent a single individual of advanced age. In addition, a fragment of a human maxilla was recovered from one of the test pits.

The test excavations show that a substantial and rich archaeological deposit exists under the modern fill and concrete floor. The greatest depth occurs from the vicinity of the iron gate extending outward for a distance of at least 20 meters. It is expected that the nature of the archaeological deposit will change substantially (probably for the worse) at the dripline.[4]

Based on these encouraging results of the three-day dig in 1977, a more thorough excavation was carried out from June 15, 1978,

---

4   Brian M. Butler, "Preliminary Archaeological Assessment of Dunbar Cave (40MT 43)," March 1977.

through September 30, 1978. The Tennessee Division of Archaeology conducted these excavations at the mouth of Dunbar Cave and they reported finding significant concentrations of well-preserved prehistoric debris under the modern fill. The artifacts discovered dated from the Archaic, Woodland, and Mississippian cultures, or from roughly 10,000 years ago until 500 years ago. Robert A. Pace and Victor Hood describe their find in archaeological terms:

A substantial and continuous sequence of occupation from Early Archaic to Mississippian is present at Dunbar Cave. Because of its abundant representation and easily defined context, the Archaic sequence appears most significant. Tentative evidence indicated the frequent, though intermittent, use of Dunbar Cave during the Early and Middle Archaic (8,000–3,000 B.C.). Early Archaic materials are represented in the lowest levels by the Kirk cluster of pro-

jectile point types. These corner-notched points strongly resemble the Cypress Creek II point type reported from the Eva site. A Beaver Lake projectile point, recovered from one of the lowest levels, suggests considerable antiquity of occupation for the cave. Also present as a majority type are narrow, lanceolate, corner-notched forms with an incurvate base resembling the Little River type defined in southwestern Kentucky. Although a Middle Archaic affiliation for the Little River point has been suggested, they occur throughout the Archaic sequence at Dunbar Cave. Identification of a Middle Archaic component is based on the presence of side and basally notched projectile points occurring in soil horizons immediately above the Kirk zones. Late Archaic manifestations are not as well defined, although a sizeable number of straight-stemmed projectile points occur. One reason for the lack of definition may be attributed to an increasing rate of

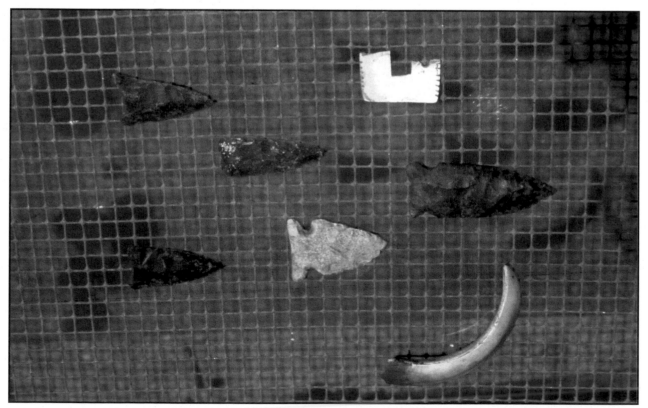

Figure 1-3 Artifacts recovered during the archaeological excavation in the Dunbar Cave Entrance. Photo courtesy of Dunbar Cave State Natural Area.

deposition of soil and rock in the upper zones of the cave concurrent with the Late Archaic through Mississippian occupations.

Diagnostic remains of the Woodland tradition are poorly represented. Early Woodland pottery is minimal and diagnostic projectile points of this period are rare. The Middle Woodland occupation is also minimally represented. Limestone-tempered, cord-marked pottery occurs in the upper layers of the midden in association with Late Woodland Hamilton projectile points. Although the Late Woodland and Mississippian are well represented, the context of some of this material is suspect.[5]

Clearly, these archeological excavations prove that Dunbar Cave was an important shelter for American Indians from the time they arrived in what is now Tennessee, nearly 10,000 years ago, until approximately 500 years ago. The early white settlers would find the cave equally as attractive as their predecessors. This will be described in Chapter 3.

Suggested Reading

Patty Jo Watson is the editor of the book *Archeology of the Mammoth Cave Area*, which was published in 1974 by the Academic Press. Since Mammoth Cave is only 80 miles northeast of Dunbar Cave, this book is a wonderful introduction to how prehistoric peoples were using caves in this area.

---

5   Robert A. Pace and Victor P. Hood, *op cit*.

Dunbar Cave

Chapter 2

Hidden Treasure

The Discovery

When you enter Dunbar Cave for the first time, one of the things that you notice on the tour is that the walls are covered with graffiti. There are some areas of the cave that are subject to occasional flooding by the River Styx that have little graffiti. Any graffiti left on these walls has been hidden by a layer of mud. But the higher and dryer passages of the cave have thousands of names and dates left by persons visiting the cave, some dating back to the early 1800s. Such graffiti can sometimes give valuable information about who visited the cave and when. But there is so much graffiti in Dunbar Cave that I hardly knew where to start when I began research for this book in 2003.

It occurred to me that it would be very helpful to have some assistance in interpreting this data, so for the research trip scheduled for January 15, 2005, I invited along local Clarksville historian Billyfrank Morrison and History Professor Joe Douglas who works at Volunteer State Community College. As Park Interpretive Specialist Amy Wallace led us through the cave, we were enjoying reading the various names and dates found on the walls. Billyfrank was able to identify some of the people who had left their names on the walls and ceilings.

About 600 feet from the Entrance, we were in the Ball Room, where a side passage leads left into the Counterfeiters Room. The names and dates were especially abundant in this area. As we were all examining different portions of the walls, Joe Douglas said: "I think this is an Indian drawing." The rest of us rushed over to see what he had found. It was a geometric drawing of one circle inside of another one. Fortunately, Joe had seen Indian drawings in other Tennessee caves and had enough knowledge about them to realize that this was very likely one, also. Further searching led to the discovery of a few more geometric drawings that Joe thought might also be Indian glyphs. These were all made by some sort of black pigment that had been drawn on the light-colored limestone wall.

That these geometric drawings might have been left by prehistoric people certainly seemed possible, since archaeological digs at the Entrance had shown that Indians had occupied that area for a long period of time. The route to the Ball Room is easy and short, so it would have been simple for the Indians to access this area with the cane torches that they used. In fact, stoke marks and charcoal had been found throughout the Historic Section of Dunbar Cave, proving that it was well-explored by these Indians. However, none of our group was archaeologists, so it would take further research to determine if these were really prehistoric art, or if perhaps some idle person on a cave tour had made them. Joe Douglas told us that he would notify archeologists at the University of Tennessee in Knoxville who had been studying similar glyphs in other Tennessee caves so that they could look at them and determine their authenticity.

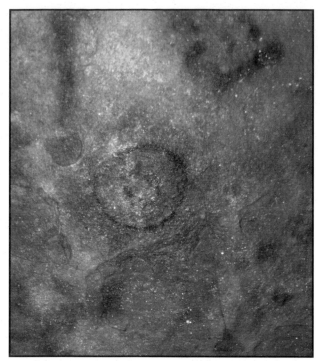

*Figure 2-1. Black pigment circle inside a circle.
Photo by Larry E. Matthews, September 26, 2006.*

The Archaeologists Arrive

Subsequent investigations by these archeologists determined that the drawings are indeed genuine, prehistoric cave drawings and etchings left on the cave walls by American Indians. According to these experts, the pictographs and petroglyphs date from the Mississippian Era, which spanned A.D. 1000 to A.D. 1600.[1]

The cave art on the walls fell into two categories: (1) pictographs, which are images that were drawn on the cave wall with charcoal, and (2) petroglyphs, which are images that were carved into the rock walls. A careful search determined that these Indian glyphs were scattered throughout the Historic Section of the cave. In all, 35 individual glyphs were located.[2]

The Glyphs

The glyphs and art work in Dunbar Cave can be classified into several basic categories:

Geometric Designs - There are several geometric designs that were drawn on the walls of the cave using charcoal or engraved into the limestone wall of the cave. These include:

Circles - These circles may be single circles or they may be concentric circles. In some cases, these circles are arranged in rows.

Circles Containing Crosses - These circles contain two lines crossing at right angles. Three glyphs fall into this category.

Circles Enclosed In Rays - Two circles are enclosed by an outer geometric figure that has eight or nine points, very much

*Figure 2-2. Unique black pigment glyph.
Photo by Larry E. Matthews, September 26, 2006.*

like a star-burst. The circle inside the starburst with eight points has two crossed

---

1   Jan F. Simek, Joseph C. Douglas, and Amy Wallace, "Ancient Cave Art at Dunbar Cave State Natural Area," *Tennessee Conservationist*, vol 73, no. 5, pp 24–26.

2   Ibid.

warrior.[3] Simek gives further information on this figure:

> This drawing is a reclining, human-like form, clearly male, with well-defined arms and legs. The head is drawn by points instead of solid fill. An ax is in the hair of the image, and a curling appendage (forelock) relates to the hair area. Claws are evident on the feet, suggesting an animal aspect. A thickening

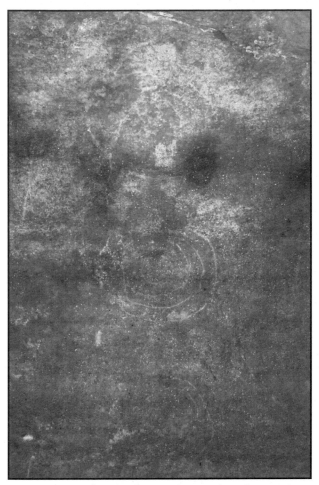

Figure 2-3. Vertical row of incised circles inside circles. Photo by Larry E. Matthews, February 28, 2006.

lines inside. The circle inside the star-burst with nine points has a smaller circle in its center containing a geometric design.

Rectangles and Squares – These are most readily identified as squares, but since they are hand-drawn, some could also be considered rectangles. One contains a line and another contains a cross.

Human Figures – There are two drawings of humans that were drawn on the walls of the cave using charcoal. They are:

Human Head – One drawing shows a human head in profile.

Reclining Human Figure – There is one remarkable drawing of a reclining human figure with a hatchet at the head. This figure shows arms, legs, and male genitalia. According to Jan F. Simek, Joe Douglas, and Amy Wallace, this figure is an anthropomorph of a Mississippian

Figure 2-4. Vertical row of black pigment circles inside circles. Photo by Larry E. Matthews, September 26, 2006.

of the figure at the waist suggests the presence of a garment on the lower torso. Warriors, a common and important

3   Ibid.

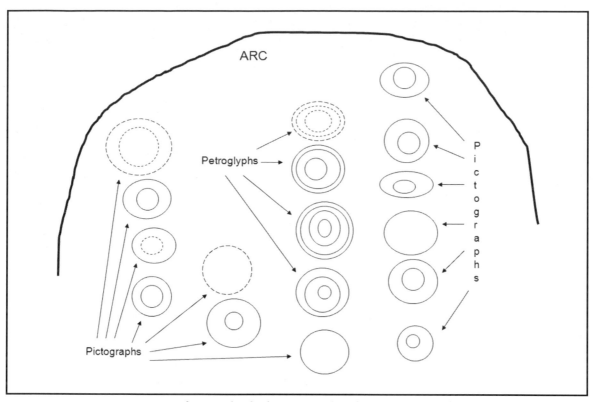

Figure 2-5. Drawing of 17 circle glyphs arranged in four vertical rows beneath an arc. Drawing © copyright by Jan Simek. Used by permission.

element in Mississippian art, are often shown with elaborate head decoration, inducing weapons, and they are often depicted in costumes that may include wings, claws, beaks, and eye decorations. Curling hair locks are also common aspects of the warrior regalia. This is a depiction of a Mississippian warrior. This fits well chronologically with the rayed circles. The image of a human transforming into another creature is an emerging theme in cave art. The creatures are inhabitants of ei-

Figure 2-6. Amy Wallace stands next to circles enclosed in rays with geometric designs in the centers. Photo by Larry E. Matthews, September 26, 2006.

ther the underworld or the sky.[4]

_____

4   Jan F. Simek, Personal communication, March 29, 2011.

Figure 2-7. Close up of two circles enclosed in rays with geometric designs in the centers. Photo © copyright by Jan Simek. Used by permission.

was made in this fashion.

What do these geometric shapes mean? They clearly had significance to the people who drew them. A valid comparison might be the Christian Cross and the Jewish Star of David. There are other geometric designs that have great religious significance to some people. A swastika could be considered as another example.

The greatest single concentration of glyphs occurs on the west wall of the Main Passage immediately past the Counterfeiters Room. This area is known as the Ball Room. Here, there is a remarkable concentration of 17 circle glyphs arranged in four vertical rows.

In addition to the designs themselves, this cave art can be further categorized by how it was drawn. There are two categories:

Petroglyphs – Petroglyphs are images that have been carved or incised into the rock. Some of the concentric circles in Dunbar Cave fall into this category.

Pictograms – Pictograms are images made by drawing on the cave wall with some type of pigment. In Dunbar Cave, the pigment used was the charcoal from the ends of the cane torches. The majority of the prehistoric cave art in Dunbar Cave

Figure 2-8. Amy Wallace stands next to the reclining warrior figure. Photo by Larry E. Matthews, February 28, 2006.

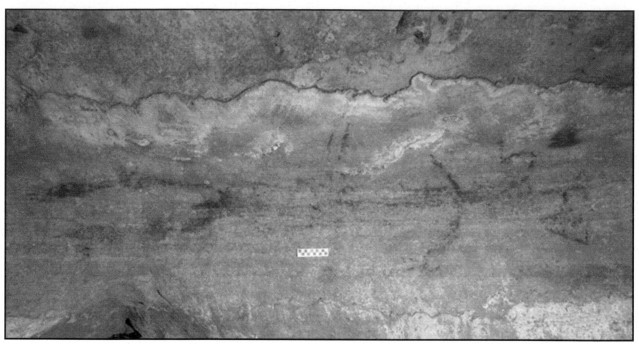

Figure 2-9. The reclining warrior figure. Photo © copyright by Alan Cressler. Used by permission.

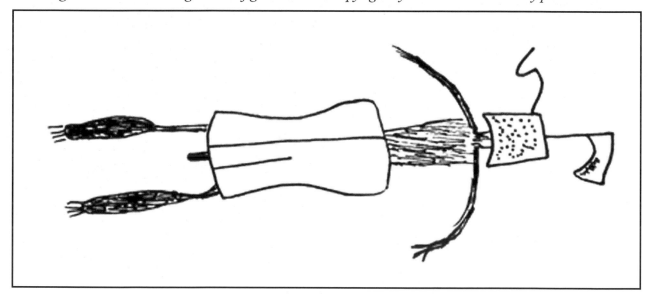

Figure 2-10. Artisitic recreation of the reclining warrior figure by Jan Simek, showing details impossible to see in a photograph. Drawing © copyright by Jan Simek. Used by permission.

These vertical rows contain from two to six circles each, placed about one foot apart. An arc passes over the top of this area, encasing all 17 glyphs inside its interior.[5] This feature is very difficult to photograph, but Jan Simek has prepared an excellent drawing. (See Figure 2-5)

The Proof

A skeptical person is bound to ask: "How can you be certain that these glyphs were drawn by Indians?" After all, an idle visitor, sometime during the 1800s, or even the 1900s, could have used some form of black pigment to draw these geometric shapes on the wall. The very reason that the glyphs were not discovered before 2005 is that they were "lost" in the thousands of pieces of graffiti drawn on the cave walls. Clearly, most of this graffiti was left on the cave walls after the arrival of European settlers. How can we be certain that these glyphs predate all

---

5   Simek, Douglas, and Wallace, op. cit.

Figure 2-11. Drawing of a profile of a human head.
Photo © copyright by Alan Cressler. Used by permission.

pieces of charcoal have been Carbon-14 dated and prove that Indians were exploring inside the cave during prehistoric times. Third, very similar Indian glyphs have been found in other Tennessee caves that have not been vandalized with graffiti, and it is a much clearer association between the glyphs, the charcoal, and the dating. Fourth, these geometric designs may appear "random" to the casual visitor, but they appear over and over in the archaeological record and they clearly had significant, perhaps religious, meanings to these prehistoric peoples. All of these pieces of evidence combined are compelling proof that these Dunbar Cave glyphs are authentic Indian drawings.

the other graffiti?

To convince ourselves that they are, indeed, genuine Indian drawings, we have several very compelling pieces of evidence. First of all, we know that Dunbar Cave was used by Indians for thousands of years. This is clearly proven by the archaeological digs that have been conducted in the entrance area. Secondly, we know that these Indians frequently went into caves, based on evidence both here at Dunbar Cave and at many other caves in Kentucky and Tennessee. Charcoal stoke marks, that were left by the Indians, are conspicuous on the walls of Dunbar Cave along with small pieces of charcoal that fell off of their cane torches. These small

The ultimate proof would be to carefully scrape some of the black pigment off the wall and subject the material to Carbon-14 dating. But these drawings are much too rare and precious to sacrifice even one for such a test. Furthermore, the intense use of Dunbar Cave for the past 200+ years may very well have contaminated the carbon pigment as other people went through the cave with a variety of torches. Some of the carbon from those torches certainly settled out on the walls and could affect any date calculation. Perhaps future testing will be able to use a smaller sample. If so, then we may be able to obtain a more direct

Figure 2-12. Black pigment rectangular glyph with interior lines.
Photo by Larry E. Matthews, September 26, 2006.

Georgia, Kentucky, Missouri, Tennessee, Virginia, West Virginia, and Wisconsin have been found to contain such art.[7] The majority of these caves, however, are located in Tennessee. All of these caves have dark zone art, where it was produced deep inside natural subterranean passageways, well beyond the reach of external light. By the Mississippian Period (A.D. 1,000 to A.D. 1,600) cave art had become common in the culture as a ceremonial expression of part of a complex religion.

date for these glyphs. But, for now, we can still be satisfied that they were, indeed, left by prehistoric Indians.

Age of the Glyphs

According to Jan Simek, Professor of Anthropology at the University of Tennessee, this remarkable cave art is from the Mississippian Era, which spanned A.D. 1000 to A.D. 1600 in North America. However, Indians have explored Dunbar Cave for a much longer time. Simek reports that cane torches left by the Indians in the cave have been dated to the Late Archaic (about 5,000 years ago), the Woodland Period (about. 2,500 years ago), and the Mississippian Period (about 700 years ago).[6]

Prehistoric cave art has been recognized in the southeastern United States since at least the 1950s, since Thomas C. Barr Jr describes such art and shows sketches in his classic 1961 book *Caves of Tennessee*. At last count, 71 caves in Alabama, Arkansas, Florida,

The Public Announcement

The Second Dunbar Cave Day was held on July 29, 2006. During that ceremony the discovery of the Indian gylphs was announced to the general public. The Tennessee State Parks system kept the discovery secret for a year and a half. During that time period, they replaced the existing gate on the cave with a new, more secure gate.

The New Cave Tour

Although Indian glyphs are known from over 50 Tennessee caves[8], there was no location where the public could go to view them. With this in mind, the state modified the existing Dunbar Cave tour so that the public could observe some of these glyphs. A protective barrier was erected to keep the public a safe distance

---

6   Jan F. Simek, Personal communication, March 27, 2011.

7   Jan F. Simek, Personal communication, March 29, 2011.

8   Jan F. Simek, Personal communication, March 29, 2011.

from the glyphs to prevent damage. Unfortunately, at the present time, all tours of Dunbar Cave have been discontinued due to the discovery of one single bat with White Nose Syndrome (WNS) in the cave.

Suggested Reading

The December, 2006, issue of the *NSS News* contains an article titled "Indian Glyphs Discovered In Dunbar Cave, Tennessee" on pages 10 through 12. The ar-

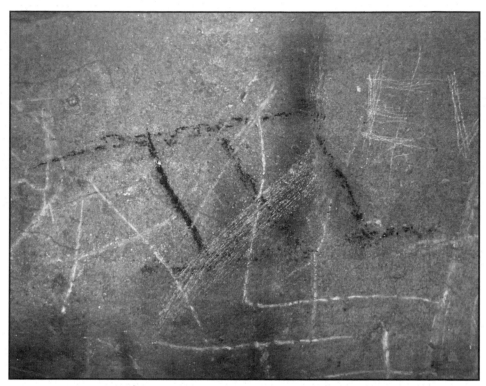

Figure 2-13. Black pigment square box with one interior line. Photo by Larry E. Matthews, September 26, 2006.

Figure 2-14. B.E. Ringonelert left his name and the date 1843 on the wall above the bear wallows, where bears once hibernated. Avis Moni lies in one of the bear wallows. Photo ©copyright by Alan Cressler, January 28, 2007. Used by permission

ticle was written by Larry E. Matthews and contains the first published photographs of the Indian glyphs including concentric circles, squares, star bursts, and the reclining warrior.

The September/October, 2007, issue of the *Tennessee Conservationist* contains an article titled "Ancient Cave Art at Dunbar Cave State Natural Area" on pages 24 through 26. The text was written by Jan F. Simek, Joseph C. Douglas, and Amy Wallace and provides a good introduction to the cave art present in Dunbar Cave. The article is accompanied by a number of excellent photographs by Alan Cressler.

In 1986 the University of Tennessee Press published *The Prehistoric Native American Art of Mud Glyph Cave*, which was edited by Charles H. Faulkner. This book describes the prehistoric art of another Tennessee cave and other sites in Tennessee, Alabama, and Georgia. If you want to learn more about Mississippian cave art, this is must reading.

Chapter 3

Early Settlers: 1790–1857

Isaac Rowland Peterson and Thomas Dunbar

The first white settlers to move into Tennessee followed one of two routes: they either crossed the Appalachian Mountains from North Carolina, or they came by flatboat up the Tennessee and Cumberland Rivers. Since crossing the mountains was very difficult and it was nearly impossible to transport goods by that route, most early settlers into Tennessee traveled by boat down the Ohio River, then up the Cumberland River into middle Tennessee. Once on the Cumberland River, they could settle on prime farmland along the river or move up the tributaries into more unclaimed land. To get to Dunbar Cave by this route, the settlers came up the Cumberland River to the approximate location of modern-day Clarksville, then they traveled up the Red River approximately 8 miles. From here they could follow a tributary of the Red River 1.5 miles upstream straight to the mouth of Dunbar Cave. The rivers and their tributaries were the early highways of Tennessee. Caver Jim Whidby wrote the following account:

> In 1790, Isaac Rowland Peterson (from Buncombe County, North Carolina) visited Idaho Springs. During his visit, he staked a claim to the springs and what are now the Dunbar Cave properties before returning to North Carolina the same year. In 1791, he returned from

North Carolina with his family, only to find that Thomas Dunbar was well established in a blockhouse (log cabin) on the land. Although the two families argued over the land rights, the Dunbars offered the Petersons shelter during the winter months of 1791–1792. According to legend, Isaac Peterson refused to stay in the blockhouse and spent much of the winter in a hollow sycamore log nearby. In 1792, following a lawsuit, Isaac Peterson received clear title to the land in a settlement of "one hundred pounds" paid to Thomas Dunbar. However, the Dunbar name remains to this day as the official name of the cave.[1]

According to legend, the mineral springs received their name, Idaho, from an American Indian. The legend stated an Indian chief, who stopped at the springs with his tribe, was named Idaho and gave the springs his name. The Indians are said to have felt renewed courage and strength after drinking the mineral waters. The word Idaho is said to mean "Jewel of the Mountains."[2]

Richard P. Geer wrote the following interesting, if somewhat speculative, account of

---

1 Jim Whidby, Idaho Springs and Dunbar Cave Resort—The Showplace of the South, *Journal of the Cumberland Spelean Association*, vol. 6, no. 1, pp 7-12.
2 *Ibid.*

the arrival of the first settlers at Dunbar Cave:

In 1790 Clarksville was a small but teeming village on the south bank of the barely navigable Red River, surrounded on all sides by a dense, untamed wilderness filled with more or less hostile Indians who rightfully viewed with alarm the invasion of the white man, changing the land in ways they neither desired nor understood. For the white man, of course, there was a wealth of land to be had here, far better land than he had found in the isolated southern mountains to the east. All one had to do was to hack a piece of land out of the wilderness. These pioneers were a hardy lot, high spirited, independent, and with a capacity for daring and hard work. Quite used to hardship, they had a willingness to apply these qualities to conquer this new and promising land.

Isaac Rowland Peterson was one such man. He left his family behind in North Carolina, struggled across the wilderness to Clarksville, and now turned his back on the security of that village. Crossing the river to its north shore, he followed a meandering stream northward into the wilderness.

Isaac soon came to a high limestone bluff with a large rockshelter high in its face. Below the rockshelter a stream gushed from a hole in the rock and plunged with a roar down through a descending, water-worn gorge to spill out into the valley below. The towering bluff was a fortress against attack from all directions save the climb up the bluff from the direction he'd come. Even the very concavity of the bluff offered protection from attack through the dense mat of forest that encroached upon its flanks. Deep within the rockshelter was a cave, which might offer further protection and could preserve his food as well. It might even contain saltpeter from which he could make the gunpowder he'd need to hunt with, and to protect his family from the savages from whom he'd no doubt have to wrench the land. And the scene from the

*Fgure 3-1. Very early photo of the Dunbar Cave Entrance showing wooden dance floor.
Postcard from the Billyfrank Morrison Collection.*

Figure 3-2. Figure of a running man painted in black. Photo by Larry E. Matthews, September 26, 2006.

rockshelter was one of serene beauty.

What more could a man ask of life in an untamed wilderness? There was abundant game; the water was pure, cold, and plentiful; and an unlimited supply of timber was available for building, for cooking, for warming his family in the winter. This was what he'd dreamed of... what he'd sought. With little further thought, Isaac Rowland Peterson staked out the land as his own, then promptly began the long trip back to North Carolina to fetch his family.

While Isaac was away, a certain Thomas Dunbar, with family in tow, chanced upon the valley. He, too, was enchanted by it. Thomas Dunbar was probably unaware of Isaac's claim. He staked out the valley as his, and quickly built a sturdy blockhouse on it to finalize his claim. So established, he moved his family into the security of the blockhouse and settled down to the hard working life of a Tennessee pioneer farmer.

With the coming of spring of 1791, Isaac reappeared with his family to take possession of his claim, but found the Dunbars well established. While possession might be a large measure of the sometimes-archaic frontier justice, a small but no less important measure was a stubborn will to fight for that which one believed to be one's own. In no way would Isaac relinquish his claim to the tract of land and, as would be expected, Thomas had no intention of giving up the property either, especially after having built his home on it. A bitter dispute raged all summer, and it is a credit to the good sense of the two men, probably under the influence of their wives, that no one was shot in the process. Such an occurrence would not have been unexpected.

With the approach of winter, the women, no doubt, were moved to Christian charity as a quick answer to the dilemma of the homeless Petersons. The Dunbars took in the Petersons. But Isaac, so the story is told, refused charity for himself and, while he expediently permitted his family to take refuge with the Dunbars, he refused to step foot in the blockhouse, spending much of the winter in a hollow hickory log.

The matter of Peterson vs. Dunbar was settled legally in 1792 at Clarksville, the seat of the embryonic Montgomery County. Isaac received a clear title to the tract in payment of "one hundred pounds" to Thomas Dunbar, according to the Montgomery County land deed register for that year, and Isaac Rowland Peterson became the first legal owner of Dunbar Cave.

It is ironic, though, that the

Figure 3-3. Portrait of a man in pastel chalk.
Photo by Larry E. Matthews, December 7, 2004.

cave was to be named forever more after the first possessor. The Dunbars held yet another distinction. Thomas Dunbar's daughter, Anne, was born in the blockhouse. Anne Dunbar was the first white girl to be born in the section.

It appears that the Dunbars and the Petersons lived harmoniously as neighbors for many years. Both families drew water from Dunbar Cave's stream. The two families probably became quite close once the problem of land ownership was

legally settled, and this was another law of the wilderness. Without it, survival would have been far more difficult than it was.[3]

More recent research by David Britton, however, suggests that the above story may be inaccurate. According to this new research, perhaps Anthony Crutcher was the first real owner of the Dunbar Cave.

3   R. P. Geer, Historical Sketch of Dunbar Cave, *The Michiana Caver*, vol. 5, no. 5, pp 37-52.

Anthony Crutcher and Robert Nelson

While conducting research for this Second Edition of the Dunbar Cave book, I ran across a typed document in the Dunbar Cave State Natural Area Office, titled: "History of Dunbar Cave Updates." Amy Wallace informs me that those were her notes from David Britton's research.[4] Attached to this one-page chronology of the cave's ownership were typed copies of early court documents for this property. They present a very different picture of the early history of the cave's ownership. This information is as follows:

1. Thomas Dunbar was likely the first modern discoverer of the cave. He was living here as early as August of 1784, which is *very* early for this area. However he did not own the land or the cave. He would today be known as a squatter.

2. Isaac Peterson had really nothing to do with the cave, thought he was a neighbor. He was here by 1787 as indicated by his 480-acre land grant and a recollection of him on a Buffalo hunt.

3. The land was originally part of a 1,000-acre grant to Anthony Crutcher, but he could not produce the paper work to prove it. He made a legal covenant with Thomas Dunbar to sell him 500 acres for which Dunbar paid upfront.

4. Robert Nelson, with proper paperwork, was able to secure rights to the land and became the first legal owner in 1792. When he acquired the land, he filed an eviction lawsuit against Dunbar, removing him from the land. Dunbar was allowed to sell the house he had built on the land. He sold the house and 8 acres to Isaac Peterson in 1792. This was near to the west of Haymarket Street.

5. Robert Nelson sold Dunbar 68 acres located to the immediate east of the cave (where the golf course is currently located) with his house likely being where the golf course club house is located today.

6. There was never any lawsuit between Peterson and Dunbar.

7. Dunbar Cave remained in Robert Nelson's possession for several years until he sold it to John Newell in 1809. Newell then sold the cave to William Tate in 1811. Tate was the third Mayor of Nashville, Tennessee, and was Mayor at the time he owned Dunbar Cave.

8. In 1814(?) Tate sold the cave to Charles Cherry. When Cherry passed away he left the property to two of his sons. Both sons were to share the cave and the springs.

9. In 1839, the oldest son, Thomas Cherry, was convicted of holding and passing counterfeit coins.

10. The cave and property were sold to make bond for Thomas Cherry, who went missing. The cave was sold to John Nick Barker in the 1840s.

Clearly, the ownership of the cave and surrounding property from 1784 until the 1840s is very complex and needs more research.

1800 to the Civil War

The mining of cave dirt for the production of saltpeter, a necessary ingredient of gunpowder, was widespread in this area during this time period. Whether, or not, Dunbar Cave was indeed mined for saltpeter and during what years this may have occurred are discussed in Chapter 4 – Saltpeter Mining.

According to Priscilla Weathersby, the Dunbar Cave property was transferred from Isaac Peterson to his grandson, also named Isaac Peterson, in 1833. Weathersby goes on to state that the next owner of the cave was John Nicholas Barker (1796–1873), "a wealthy land and slave owner, a busy farmer, and family man." But the exact date is not given for the transfer of the property. Barker may have acquired the property in 1843. Mr. Barker owned the Dunbar Cave property until his death in 1873. According to a diary kept by

4   Amy Wallace, personal communication, April 25, 2011.

Figure 3-4. Face drawn on a rock. Photo by Larry E. Matthews, September 26, 2006.

Mr. Barker, there was a large party and bran[5] dance held at Dunbar Cave on July 7, 1844. This is the first known account of the cave being used as a social gathering place. This was the day when Bryant Peterson fell into a ravine inside the cave and the story of "Peterson's Leap" originated.[6] A story since located in the *Tobacco Leaf Chronicle* newspaper and reprinted below, confirms this belief.

5   These early dances were called bran dances because bran was spread to a depth of several inches on the ground. Some people have assumed this to be a misspelling for *barn*, but that is incorrect.

6   This account is from a 25-page history of "Dunbar's Cave" that is included in one of the notebooks on file at the Dunbar Cave State Natural Area Office. It is titled: "Dunbar Cave State Natural Area Research Project, Theme 4: History", by Deborah Lingle Gillis. This particular 25-page history ends with the note: Signed by Priscilla Weathersby, 1981.

Petersons Leap - July 7, 1844

Jesse Glass discovered the following first-hand account of the story of Peterson's Leap in the Wednesday, July 29, 1874, edition of the *Tobacco Leaf Chronicle*:

Dunbars Cave. The Adventurer of Peterson's Leap Visits Scene of his Miraculous Escape Thirty years Ago. A Joyous Sabbath School Picnic.

Last Wednesday was a day most pleasantly spent by the Baptist Sunday School at Dunbar's Cave. The invigorating air from the cave seemed to give new life and animation to all, old folks became young, and it was pleasing to see all mingling in childish play, engaging in such innocent amusements as would interest the children.

The attendance was full, and a more elegant dinner we have never seen on such

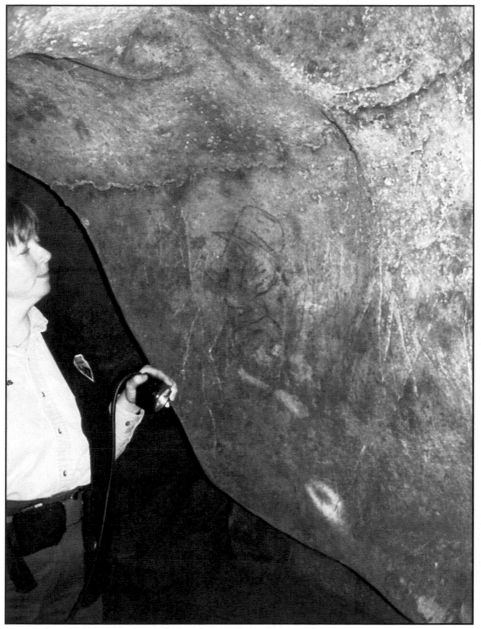

Figure 3-5. Amy Wallace by a drawing of a man wearing a hat. Photo by Larry E. Matthews, December 7, 2004.

find great comfort in a day spent there, where there is neither summer or winter, but all the time a pleasant invigorating atmosphere. What a pity the place can't be fitted up for the accommodation of guests. The Cave in connection with Idaho Springs half a mile below, with good buildings, would form unsurpassed attractions.

We had the pleasure of meeting at the cave Bryant F. Peterson, ESQ., the gentleman whose name will be perpetuated in connection with the cave during time. He was there reviewing the scene of his daring adventure, know as "Peterson's Leap," the story of which, and his miraculous escape, has been repeated over and over again to every visitor of the cave and will continue to be told for ages to come. That we might hear it correctly and learn all of the circumstances we asked Mr. Peterson to repeat it afresh.

It was 30 years ago (in July) he said a crowd of people, several hundred, had gathered at the cave for a bran dance. A party of young men and boys determined to explore the cave and requested him to serve as guide, which he consented to do, being familiar with all of the routes from

occasions. The school went out in a body about nine o-clock and remained until five in the evening.

It was a day of real refreshing recreation and pleasure, and all were joyful and happy. No place on earth, or rather in the earth, equals Dunbar's Cave for such occasions, recreation, and pleasure.

Persons who are depressed and worn down from heat and exhaustion may go there for relief and have their energies and vigor renewed. The invalid will often

frequent visits. In fact he was raised within a few hundred yards of the place, and was often called upon for such service. When the party arrived at the precipice now known as Peterson's Leap, Mr. Peterson approached near the brink for the purpose of pointing out a singular formation in the rock overhead, which excited so much interest that the boys gathered close up, and one of them slipped or stumbled and pushed him over. He fell 35 feet, perpendicularly, striking on the back of his neck, or rather shoulders on a sloping bank of dirt. He was for some minutes stunned by the fall but did not feel seriously hurt. The question of getting out was a serious one. The boys proposed to tie their handkerchiefs together and make a rope. He told them that he was afraid to risk it, but could pull up by their

suspenders if tied together. In a minute every fellow had off his suspenders and were making a rope, when one of the boys discovered a grapevine, which doubtless had been used to measure the depth of the caverns. The vine was let down, and he fastened on to it and (he was) pulled up. In a minute after reaching the top he fainted and was for several days unconscious of anything and his recovery was for a long time a matter of grave doubt. Prompt medical attention and good nursing was all that saved him, and no doubt had he remained in that horrible pit 10 minutes longer he would never have been extricated alive. Mr. Peterson has lost one arm, with that exception, he is a portly, fine looking gentleman, a man of fine address, and evidently a good lawyer. His home is Carthage, Illinois. He is here with

Figure 3-6. A hand points the way "OUT". Photo by Larry E. Matthews, December 7, 2004.

his family on a visit to his friends in this county, and will probably remain until the first of September.

Mr. Peterson says he can discover little, or no change about the cave. The growth and scenery around is the same, especially the large grapevine at the mouth looks just as it did 30 years ago.

There are two routes in the cave leading to Peterson's Leap, approaching from different directions. Parties will see the place to much better advantage by dividing in two companies and going up to it on both sides. The light will show from each side, and by throwing a torch may judge of the depth, besides the light is beautiful.[7]

This newspaper account is very informative on three levels. First of all, it tells us that large gatherings and bran dances were already being held at Dunbar Cave by 1844. Secondly, it tells us that guides were already leading tours of the cave by 1844. And, thirdly, it gives accurate details of Peterson's accident and the date (July 7, 1844) that it occurred.

A Summer Picnic, August 5, 1851

The following account of a summer picnic on the grounds at Dunbar Cave was printed in the Clarksville *Jeffersonian* in the Wednesday, August 9, 1851, edition. This information was located and provided by Jesse Glass.

We had the pleasure of being present at the picnic at Dunbar's Cave on Saturday, which was to be the most pleasant and agreeable affair that has ever come off at that romantic and beautiful spot. It has for a long while been our custom to resort to this cave as being by far the most attractive retreat anywhere in the vicinity of Clarksville, but to visit it under a circumstance so peculiar as those of Saturday, will long be a cherished memory with us. Several ladies

were present, among others, who by their sparkling wit contributed largely to the spirit and humor of the day. However, that which affected us in a peculiar way, was two or three beautiful duets, sung by two young ladies selected for this purpose by their companions. It produced within the silent and gloomy cave, an effect the most delightful and charming. We never heard gems of music rendered so beautifully. The scene, the hour, in fact everything seemed to conspire to heighten our emotions—the feeble glare of the lights as they struggled with the utter darkness that rose before us, the sepulchral echo as it died away in some secret cavern, fell upon the ear and heart with a strange yet thrilling power. We thought as we stood apart from the charming and gifted singers, and listened to the wild melody of their song, that we were actually in some old cathedral vast and dim, among whose solemn arches had gathered the grim and dusky shadows of a thousand years; and had a slow and mournful chant been sung, the illusion would have been complete, and imagination would have thrown around us the black interior of a convent. We never can forget the impression within the cave on that day. The cheek of beauty and loveliness as it grew deadly pale from the smothered light or perchance from some secret fear, made us look upon them as a band of exiled spirits from some serene and bright abode.

After the young ladies had finished their songs, all sought once more the "merry sunshine," and partook of a repast, spread upon the ground in lavishing profusion, consisting of all the delicacies and luxuries calculated to tickle the palate on such occasions. But we have said more than we intended, and we "round to" by expressing the hope that the boys will get up just such another affair before the summer is ended.

Jesse Glass notes that this newspaper article is unsigned, but he suggests that it was

---

7   Jesse Glass, Personal communication, July 26, 2009.

Figure 3-7. A head with a nose like a woodpecker drawn on the cave wall.
Photo © copyright by Alan Cressler. Used by permission.

most probably written by Henry Faxon.[8]

According to the research by Deborah Lingle Gillis (June, 1993), the owner(s) of Dunbar Cave from Barker's death in 1873 until 1879 are not known, but in 1879 Charles Warfield purchased the property from Chancery Court. The property changed hands again in 1882 when it was purchased by J.M. Rice, C.P. Warfield, and J.P. Gracey. It is this last group of owners who began to develop Dunbar Cave into a tourist attraction. This will be discussed in Chapter 5: Early Commercialization.

8   Jesse Glass, Personal Communication, November 30, 2008.

Chapter 4

Saltpeter Mining

Saltpeter Mining West of the Appalachian Mountains

Daniel Boone followed the Warriors Path through the Cumberland Gap in 1775, paving the way for Caucasian settlers, along with their African slaves, to move west into Kentucky and Tennessee. According to historical records, Daniel Boone was the first person to make gunpowder in Kentucky. Professional geologist and spelean historian Angelo I. George provides the following information:

> On the Kentucky frontier, far inland from east coast markets, pioneer settlers, backwoodsmen, and explorers made do from what the land could provide. Two indispensable commodities needed on the western frontier were salt and gunpowder. Stockades and settlements sprang up along game trails near intersections with salt licks and springs. Caves and sandstone rock shelters were excavated and processed for saltpeter, first for personal use. By 1802, saltpeter and gunpowder manufacturing were fairly well established industries in Kentucky and neighboring states. The most significant sites were Mammoth Cave and Great Saltpeter Cave in Kentucky, Wyandotte Cave in Indiana, and Big Bone Cave in Tennessee. Collectively, the historic artifacts and documents associated with these caves present a coherent picture of life and working conditions in saltpeter mines.
>
> At first only a few people on the frontier knew the secrets of saltpeter and gunpowder manufacturing. Backwoods icon Daniel Boone and his contemporary, an African American, Monk Estill, are the first on record to have made gunpowder in Kentucky. As the region became more settled and industry developed, individuals such as John James DuFour, Fleming Gatewood Sr, Levi Brashear, Archibald Miller Sr, and African American "Free" Frank McWorter made saltpeter in commercial quantities. At the height of the industry, there were hundreds of these "strolling chemists" as they were termed in 1809 by Lexington entrepreneur Samuel Brown, MD.[1]

George goes on to note that saltpeter mining began at Mammoth Cave in 1798. Dunbar Cave is only 75 miles from Mammoth Cave, so it would be reasonable to expect that saltpeter mining could have begun in the Clarksville area at about the same time.

Saltpeter Mining In Dunbar Cave

Jim Whidby reports that: "As early as 1847, calcium nitrate was mined from Dunbar Cave. It was shipped to Sycamore Creek Powder Mill near Ashland City, Tennessee, to be refined into saltpeter for gunpowder."[2]

---

1 Angelo I. George, *Mammoth Cave Saltpeter Works*, H.M.I. Press, 2005, pp 2–3, 5.

2 Jim Whidby, "Idaho Springs and Dunbar Cave

Figure 4-1. Bert Denton stands next to a double-V saltpeter vat located in Big Bone Cave, Van Buren County, Tennessee. V-shaped saltpeter vats might have been used in Dunbar Cave. Photo by Tom Barr, about 1953.

The *Tennessee Encyclopedia of History and Culture* states that: "During the Mexican War saltpeter for gunpowder was mined at the [Dunbar] cave." Saltpeter mining was frequently a small-scale operation that allowed a cave owner to supplement his farm income. However, a few caves, such as nearby Mammoth Cave, had very large mining operations. At the present time, we have no information on the size of the mining operation at Dunbar Cave.

A description of Dunbar Cave in a June 11, 1881, newspaper mentions the remains of saltpeter mining that were still visible at that time:

> On entering the cave you are met with a strong draft of air. The cave at this point is about 30 feet wide with the roof or top

20 feet high, growing broader and higher as you advance. After a gradual descent for a distance of 50 or 60 yards, you come to a large pool of running water where turning sharply to the right, up a short hill, you enter the "wheel barrow track" which leads to the "Ball Room," a spacious hall with high ceilings and a dry pleasant atmosphere. Here are the ruins of the old saltpeter works.

So, the saltpeter mining operation was located in the Ball Room. This would make perfect sense, since this is a large, convenient location, centrally located in the cave. No remains of the "old saltpeter works" remain today. They were probably removed during the trail building in the cave.

Cave saltpeter is a naturally occurring mineral, $Ca(NO_3)_2$, that accumulates in the soil of caves. Since it is highly soluble in water, this

Resort—The Showplace of the South," *Journal of the Cumberland Spelean Association*, vol 6, no. 1, pp 7–12.

mineral accumulates in significant quantities only in the drier portions of caves. Saltpeter was mined from caves in Tennessee, Virginia, and Kentucky as early as the War of 1812. Nearby Mammoth Cave, Kentucky, was an especially important source of saltpeter for gunpowder during the War of 1812.

Clearly, a great deal of dirt has been excavated in the Historic Section of Dunbar Cave. However, there is no way to tell how much of this dirt was excavated and processed for saltpeter and how much was excavated to facilitate the commercialization of the cave. Since Dunbar Cave tends to be somewhat damp, it is probable that the yield of saltpeter was low, or even unprofitable, which would explain why this well-known cave may not have been mined during the War of 1812. Furthermore, the northern portion of middle Tennessee was quickly occupied by Federal troops after the Civil War broke out, so it was probably never an option for the cave to be mined during the

Civil War. There is little evidence for any caves in Tennessee being mined for saltpeter to make gunpowder after the Civil War.

In order to mine saltpeter from Dunbar Cave, it would have been necessary for the miners to construct one or more leaching vats. These would have been constructed from local wood using simple hand tools. Over a hundred Tennessee caves are known to have been mined for saltpeter and the leaching vats are still present in some of those caves. Therefore, we know that there were two commonly used types of saltpeter vats: square and V-shaped. Photographs of a variety of these vats can be found in *Descriptions of Tennessee Caves*.[3] At some caves, the leaching vats were located outside or at the entrance. After the mining operations had ceased, weathering soon destroyed these vats.

Once the leaching vats had been built,

---

3   Larry E. Matthews, *Descriptions of Tennessee Caves*, Tennessee Division of Geology, 1971.

Figure 4-2. Daniel Wright next to a square saltpeter vat located in a cave in Van Buren County, Tennessee. Square-shaped saltpeter vats are the other major style of leaching vats that might have been used in Dunbar Cave. Photo by Bob Biddix, August 19, 2000.

the miners would excavate saltpeter-bearing dirt and place it into the vat until the vat was nearly full. Normally, the vat was lined with straw to help prevent dirt and water from leaking through the cracks between the wooden planks. Water would have been collected from the River Styx and slowly poured over the dirt in the vats. The miners would then hand-mix the dirt and water to the consistency of mud. The water would dissolve the highly-soluble saltpeter as it worked its way downward. This saltpeter-enriched water, called *liquor*, would be collected in wooden water troughs at the base of each leaching vat.

The next process was to pour this liquor through another smaller leaching vat which was full of oak and hickory ashes. These ashes are rich in potassium and this converts the cave saltpeter, $Ca(NO_3)_2$, into potassium nitrate, K_2NO_3, which is the type of saltpeter necessary for the manufacture of gunpowder. This process also caused some other unwanted minerals in the liquor to precipitate out, so that they could be easily removed by straining.

The resulting liquid was then placed in large iron kettles, where it was carefully boiled until much of the water evaporated, leaving behind crystals of saltpeter. These crystals were carefully removed, dried, and packed in bags, so that they could be shipped to the powder mill. At the powder mill, the saltpeter crystals were ground with charcoal and sulphur to make gunpowder. Apparently, the saltpeter produced at Dunbar Cave was shipped to the Sycamore Creek Powder Mill near Ashland City, Tennessee.

Other Saltpeter Mines In Montgomery County

Unfortunately, there is very little evidence in Dunbar Cave to prove the size, or even the location, of the mining activity. This is due to the large amount of dirt that was removed to develop the cave into a commercial attraction. However, it is possible to look at the evidence of saltpeter mining in other caves in Montgomery County in order to obtain some information on both the time that this activity took place

and the size of the various operations.

According to Thomas C. Barr Jr, Bellamy Cave was intensively mined for nitrates during the Civil War and the rotten remains of several old niter hoppers could be seen in both branches of the cave in 1961.[4] Barr also reports that Cooper Creek Cave shows evidence of a large amount of digging and that a great number of casts of the niter vats still remained in 1961.[5] What is so significant about these two caves is that the evidence of the saltpeter mining was still intact in 1961 and indicated rather large and significant operations.

Further research on saltpeter mining in Bellamy Cave was conducted by cave historian Joseph C. Douglas from 2006 to 2008. Douglas reports finding the remains of two dozen saltpeter leaching vats inside the cave, along with trails and steps built by the miners. He also noted extensive dirt removal, along with mattock marks, piled rocks, and possible tally marks. This clearly confirms Barr's report that Bellamy Cave was a large-scale saltpeter mine. However, Douglas raises some interesting questions about the date that this activity took place. As he notes, Clarksville and Montgomery County fell to the Union forces early during the Civil War. Nearby Fort Donelson was overrun by Ulysses S. Grant in February, 1862. Therefore, Douglas suggests that Bellamy Cave was more likely mined sometime between 1807 and 1815, the time period of what today is referred to as the War of 1812. Douglas was even able to locate shipping records from 1807 detailing cargoes of gunpowder being shipped from Clarksville to Natchez and New Orleans. This doesn't prove that the saltpeter to make the gunpowder came from Bellamy Cave, but it does document that Clarksville was a source of gunpowder during that time period. Perhaps even more important was the discovery of an article in the September 8, 1877, Clarksville *Weekly Chronicle* that

---

4   Thomas C. Barr Jr, *Caves of Tennessee*, Tennessee Division of Geology, Bulletin 64, pp 326–328.
5   Thomas C. Barr, Jr, *Caves of Tennessee*, Tennessee Division of Geology, Bulletin 64, p 330.

reports that Bellamy Cave was mined in 1812 by a Mr B. Bayless and others.[6] This research by Douglas does not conclusively prove that no mining occurred in Bellamy Cave at the beginning of the Civil War, but clearly indicates that the bulk of the mining occurred much earlier. This, in turn, raises significant questions about when saltpeter was mined in Dunbar Cave. If Dunbar Cave was mined for saltpeter, mining might have occurred during the War of 1812 era, rather than during the Mexican War. Or, it is even possible it could have been mined during both periods.

If saltpeter was mined at Dunbar Cave during the Mexican War, this conflict may have provided an increased demand for gunpowder, causing the miners to look for new sources of saltpeter. Increased demand always resulted in a higher price for saltpeter.

Cave Saltpeter as a Tobacco Fertilizer

The January 27, 1886, Wednesday morning edition of the *Daily American* (Nashville, Tennessee) ran an article on "Cave Saltpeter as a Tobacco Fertilizer." This article goes on to state:

> Clarksville, Jan. 25 – Mr. Joshua Rice has had a force of men at work in Dunbar's Cave during all of the extreme cold weather. The temperature is quite pleasant in the cave now, and men at work perspire freely, and can put in full time. Some suppose that Mr. Rice is digging after a rich treasure thought to have been concealed in the cave many years ago by a notorious character named Dick Cherry. Interviewing Mr. Rice on the subject we learned differently. True, he is after a fortune, but not Mr. Cherry's ill-gotten gains, and there is no doubt as to the find of a rich deposit that will prove a great mine of wealth. This deposit is the most valu-
>
> able tobacco fertilizer yet discovered. It has not been analyzed, but it is a natural deposit. Its component parts, principally saltpeter and lime, and properly mixed for use just as it comes from the cave. Mr. Rice brought out a considerable quantity last spring, giving it a test by experiment, using very small portions in tobacco hills. The result was an improvement of 300 percent. In other words, every hill of tobacco having this fertilizer was worth three grown without it in the same field. Some of it was used on an old road, without a particle of soil, taken into the field for cultivation and made very good tobacco, while plants set without the fertilizer made nothing. The object is not to advertise a new-fangled fertilized for sale. It is not for sale. Honorable C.P. Warfield and Mr J.M. Rice, owners of the cave, both have farms near by, and they propose using this product freely this year, testing it thoroughly on everything planted, and will give their neighbors and friends samples for trial. The men employed are engaged in what is known as the "wheelbarrow track," so that a wagon can drive into the cave half a mile or more to this saltpeter bed and haul it out to be used on the farm as fertilizer.
>
> The proprietors think that this deposit will amply pay for cutting the way in through this crawling place. Another object is to improve the cave for public entertainment—open a wagon way one mile or more, make a smooth, wide gangway, level up Independence Hall, smooth the surface of the Ball Room, and arrange Great Relief Hall, especially for the meetings of the Press Association. They are after killing two birds with one stone, getting the fertilizer and fitting the cave up ready for the Farmers' Grand Reunion and Stock Show and meeting of the Press Association next summer.
>
> In the meantime an engine with electric outfit will be purchased for illuminating the cave throughout, which will furnish an indescribable scene of grandeur

6   Joseph C. Douglas, "A Note on the History and Material Culture of Bellamy Cave, Tennessee," *The Journal of Spelean History*, Vol 43, No. 2, Issue 136 (July–December, 2009), pp 4–10.

for the crowd of people who usually attend the farmers' reunions at this place. When brilliantly lighted with electricity through all of its caverns, byways, and magnificent halls, some 3 or 4 miles underground, it will furnish a day's entertainment and ample accommodations for 20,000 people.

It certainly appears that the cave's owners, the Honorable C.P. Warfield and Mr Joshua M. Rice, were hard at work to improve Dunbar Cave for tourists by digging out the low spots to make comfortable walking trails. It is also interesting to note that many features in the cave already had their current names. If the plans for the electric lights materialized, then Dunbar Cave was certainly one of the first caves in the country to have them. Also, according to this article, the cave was already being used for large conventions. The above article mentions having "ample accommodations for 20,000 people." Clearly Dunbar Cave was already a major commercial operation by 1886. Whether or not the cave dirt was really that good as a fertilizer is another question. Was this merely a scheme on the owners' part to get the waste dirt hauled off for free? Or did they think they would be able to sell the dirt? We will probably never know.

Chapter 5

Early Commercialization: 1858 – 1899

According to the *Goodspeed's History of Tennessee*: "One of the earliest settlements in the county was made at Idaho Springs, the waters of which were early sought for their medical virtues. No effort was made to entertain boarders there, however, until 1858, when a number of cabins were built with a view of erecting a large hotel. The Civil War coming on caused an abandonment of the project, and the property lay comparatively unoccupied until the present incumbent, J.A. Tate, purchased it in 1879 and employed a skillful Irishman to assist him in opening up the long-slumbering streams. Finding traces of iron water he followed them to their source and discovered fine chalybeate waters, till then entirely unknown here. He also found the strongest sulfur water, dark in appearance, and, discovering a change in the taste of the water at different times, inferred that there were several waters blended. After much labor and expense the different waters were entirely separated, the strongest water becoming as clear as crystal. There are now three distinct varieties of sulfur water, white, red, and black, as well as a strong chalybeate and a silvery sparkling water of little mineral taste, making as fine a collection of mineral waters as can be found in any state. Rheumatism, dyspepsia, chronic diarrhea, all cutaneous and blood diseases, diabetes, and gravel yield promptly to their influence, and in consequence this is becoming an exceedingly popular resort, especially in connection with Dunbar's Cave."[1]

Because it was a widely held belief that drinking spring waters could cure many common ailments, people suffering from various maladies would spend a weekend, or longer, at a small hotel near such springs and drink several glasses of the various spring waters each day. Sulfur water was held in especially high esteem. Perhaps the belief was that the worse the water tasted, the more curative powers that it held. In all likelihood, the peace and quite and natural beauty of these resorts gave people the vacation time they needed and they really did feel much better by the time they left and went home.

A Confederate Soldier in Dunbar Cave in 1863

An article by David Britton in the February 19, 2008, *Cumberland Lore* reports on the following inscription on the walls of Dunbar Cave:

> Captain G. Holt
> Southern 3d Ky Reg
> August 1863
> Beware Yankey

The inscription itself is in black pigment on the tan cave wall and, like many such inscriptions, is somewhat hard to decipher. Further research by David Britton revealed that Cap-

---

1 The Goodspeed Histories of Montgomery, Robertson, Humphreys, Stewart, Dickson, Cheatham, Houston Counties of Tennessee, 1886.

tain Gustavus Holt was a Confederate soldier from Murray, Kentucky. The question arises, how did Holt enter Dunbar Cave in August of 1863 when Clarksville was under Union occupation? His theory is that Holt entered in civilian clothes. This would have been quite risky, since any enemy soldier captured out of uniform would be assumed to be a spy and would be hanged. Britton provides the following additional information:

Of all the signatures in the cave, one of the more interesting ones was written by a Confederate soldier in the year 1863— right in the middle of the Civil War. This signature is located a couple dozen feet before "Peterson's Leap" on the left-hand wall near to the ground inside Dunbar Cave. Through research, this signature has been identified as being written by Captain Gustavus Adolphus Christian Holt of the 3rd KY (Mounted) Infantry Regiment Co. H.

Gustavus Adolphus Christian Holt was born on March 2, 1840, in Salem, Livingston Co., Kentucky. He was the son of James Patton Holt and Julia K. Hodge, the father being a well-respected doctor

in Kentucky. He received his early education in Henderson and Murray, Kentucky and graduated from the State University at Louisville with a degree in Law in 1859. He commenced opening a law office thereafter in Murray, but the effort was soon disrupted by the opening of the American Civil War.

Previous to the war, Holt had been a state militia inspector and when the war began, Holt joined the cause and assisted in the organization of Company H, Third Kentucky Regiment of Infantry (Confederate), and was elected Captain of the company. This regiment was organized at Camp Boone in northeastern Montgomery County, only a couple of miles from the current Clarksville city limits. This regiment participated in the Battle of Shiloh, in which most of the officers lost their lives and then took part in other engagements such as Corinth, Fort Pillow, Johnsonville, the Franklin-Nashville campaign, Champion Hill, and Vicksburg. On March 25, 1864, he was made a Lieutenant Colonel and by the November following, he was promoted to the rank of Colonel. Later, as

Figure 5-1. Dunbar's cabins.
Postcard from the Billyfrank Morrison Collection.

Figure 5-2. Dunbar Cave Entrance, Clarksville, Tennessee.
Postcard from the Billyfrank Morrison Collection.

a part of General Nathan Bedford Forrest's command, he surrendered at Gainesville, Alabama. At the close of the war after his regiment was mustered out on May 4, 1865, he returned to Murrray, Kentucky to practice law.

Record-wise, Holt's regiment "disappears" after August 12, 1863 at Trenton, Tennessee and is not seen again until March 25, 1864 in Paducah, Kentucky when the regiment was mounted and eventually placed under Forrest's command. It is not exactly known what happened during this interval, but there are several possibilities. After Trenton, we do know that the regiment was short of the larger portion of their men, due to engagements such as Shiloh and Vicksburg, so it is possible that during this time, they were on a recruiting mission. We also know that after Vicksburg, there were several deserters from Johnson's and Bragg's armies all throughout Tennessee, and Holt's regiment was in Johnson's army. Either of these scenarios could possibly explain why Holt was at Dunbar Cave, near Clarksville, in August of 1863. But Clarksville was under Union occupa-

tion—a dangerous place for an officer of a Confederate regiment.

It is likely that whether Captain Holt was on a recruiting mission or simply on furlough, he was in civilian clothes and since he was from out of town, no one was any the wiser. According to a diary from this time, "Colonel Bruce is giving a grand picnic at Dunbar's Cave," the date being August 18, 1863. In speculation, it is possible that Captain Holt was at this picnic in the midst of several hundred Union soldiers, and dressed as a civilian. Perhaps he went in the cave and left his "message" to the Yankees right under their noses! It is also interesting to note, that after close inspection of the writing, it appears to have been done with a small piece of charcoal, as the letters at the beginning of the message are thick and dark and gradually fade with the last letters being barely legible.

There is a lot of speculation here but the facts are: There was a Confederate officer at Dunbar Cave in August of 1863, during Union occupation, and he left a message on the cave wall. What the particulars were/are we may never know, but in the meantime, we can do the best we can

Figure 5-3. "River Sticks."
Postcard from the Billyfrank Morrison Collection.

with what we have.[2]

After the Civil War, the Ku Klux Klan was present in the area. The book *Historic Clarksville – The Bicentennial Story* reports that the

Klan met regularly at Dunbar Cave and in the basement of Stewart College in 1868.[3]

* * * * * * * * * *

2   David Britton, "*Beware Yankeys!*," Cumberland Lore, February 19, 2008, Page 4.

3   Charles M. Waters, "Historic Clarksville – The Bicentennial Story 1784-1984," Historic Clarksville Publishing Company, 1983.

Figure 5-4. Solomons Pool.
Postcard from the Billyfrank Morrison Collection.

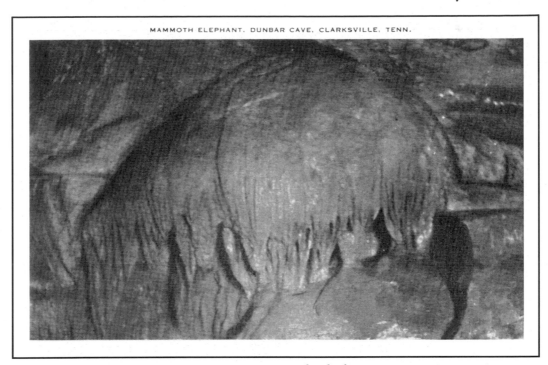

MAMMOTH ELEPHANT. DUNBAR CAVE. CLARKSVILLE. TENN.

Figure 5-5. Mammoth Elephant.
Postcard from the Billyfrank Morrison Collection.

According to the May 18, 1876, edition of the Nashville *Daily American*, Dunbar Cave was being used as a meeting place prior to Tate's purchase of the property:

Will Meet at Dunbar's Cave

Guthrie, Ky., May 17.—(Special) The annual meeting of the Montgomery County Agricultural Association will be held at Dunbar's Cave, near Clarksville, on Tuesday and Wednesday, Aug. 24 and 25. The attention of all members of "The Great Relief Press Association" is called to this meeting. All members are requested to publish this notice, and give it general circulation through the press in Tennessee and Kentucky. It will be the finest stock show ever held in Tennessee.

Frank M. Duffy,
Secretary of the
Press Association

In 1878 a newspaper dated August 27 contained a report by a Port Royal correspondent about a trip to Dunbar Cave. "They left at dawn in their buggies, took a man servant, two violins, a croquet set, provisions (including fruit) and a large brown blanket. They went inside the cave with flickering torches and traveled over a mile, looked down at that horrible cavern known as Peterson's Leap, and climbed rugged Rock Mountain and stood beside the dismal creek. Seven of our party came out by Mrs. Dancey's and spent the night under her hospitable roof. Another 1878 newspaper states: "What would Clarksville do without Dunbar Cave? Parties visit every Day."

The first mention of a hotel on the property was reported as follows: "In 1879, Professor James A. Tate of Hickory Wild Academy purchased the springs and cave property. Tate employed a skillful Irishman to assist him in building the first Idaho Springs Hotel. The hotel offered nice accommodations and provided excellent meals, and some families built summer homes around the resort grounds. Mineral waters were also bottled and sold for many years."[4] However, on June 21, 1879, the

---

4   Jim Whidby, "Idaho Springs and Dunbar Cave Resort—The Showplace of the South," *Journal of the Cumberland Spelean Association*, vol. 6, no. 1, pp 7-12.

INDEPENDENCE HALL, DUNBAR'S CAVE, CLARKSVILLE, TENN.

Figure 5-6. Independence Hall.
Postcard from the Billyfrank Morrison Collection.

newspaper says that Charlie Warfield "bought the Cave" at a Chancery sale. In all likelihood, Tate owned Idaho Springs and Warfield owned Dunbar Cave.

In 1880 a "house for the accommodation of the ladies" was built at Dunbar Cave and the "ground in front of the cave" was enlarged to provide "room for fifty couples to dance at one time."[5]

A June 11, 1881, newspaper article describes the natural entrance to Dunbar Cave:

> The scenery at the entrance of the cave is grand, large clifts (sic.) of limestone jutting out over the mouth and rising to the height of 100 feet, crowned with forest trees and shrubbery through which immense grape vines wind and twist in fantastic shapes, form a natural bower....
>
> From just below the entrance, through a deer gorge, flows a bold, brawling stream which has buttered its waters in the sacred chambers of the cave. Where this stream breaks forth, and some distance below, are

great piles of loose boulders which are covered with a profusion of beautiful ferns, mosses, and wild flowers....

> At the entrance proper is a wide open space of 75 feet in length by 40 in width entirely covered by the overhanging stones. The distance from the floor to the roof being about 30 feet. This floor has been leveled off and covered with saw dust and is used as a dancing ground, there being ample room for a dozen seats.[6]

This gives us some good idea of what the natural entrance looked like before the modern development that we see today. This newspaper article goes on to describe the cave tour:

> On entering the cave you are met with a strong draft of air. The cave at this point is about 30 feet wide with the roof or top 20 feet high, growing broader and higher as you advance. After a gradual descent

---

5   Charles M. Waters, op. cit.

6   This text is from the notes prepared by Priscilla Weathersby and dated 1981 that is on file at the Dunbar Cave State Natural Area Office.

for a distance of 50 or 60 yards, you come to a large pool of running water where turning sharply to the right, up a short hill, you enter the Wheel Barrow Track which leads to the Ball Room, a spacious hall with high ceilings and a dry pleasant atmosphere. Here are the ruins of the old Saltpeter works. Then comes a succession of rooms of different sizes: the Star Chamber, Pebble Chamber, Independence Hall, and various others, in all of which are found crystal formations, stalagmites, etc. These chambers are connected by passages, the roofs or ceilings of which come so low

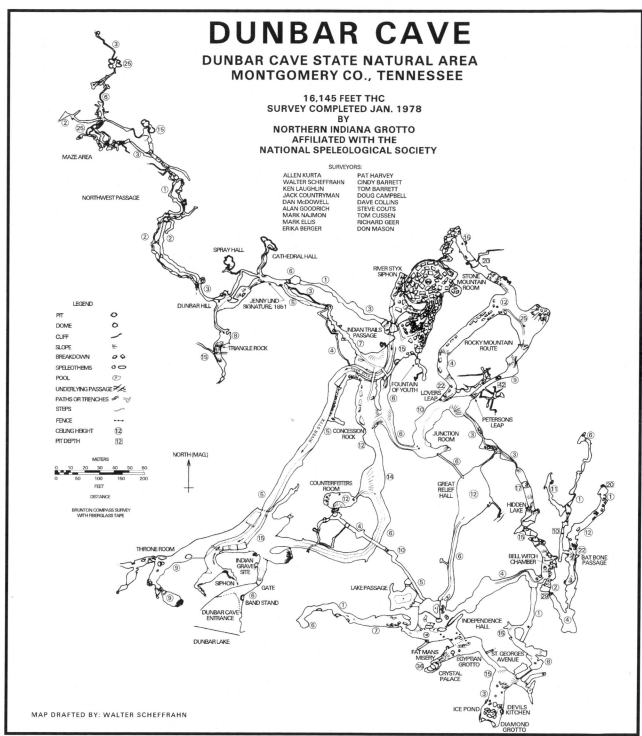

Figure 5-7. Map of the Historic Section of Dunbar Cave.

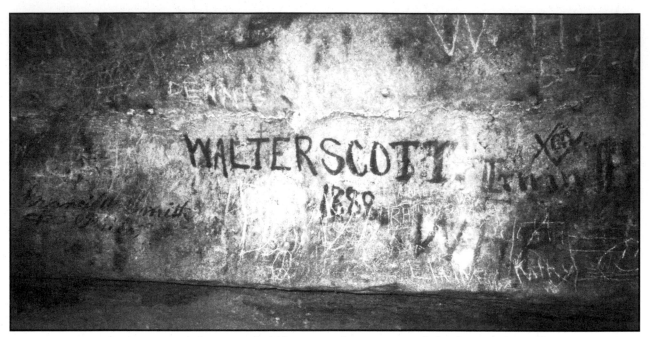

*Figure 5-8. Early visitors loved to write their name and the date on the wall.
Photo by Billyfrank Morrison.*

as to induce very humble attitudes. These passages are called the Crawling Places and Fat Man's Misery, etc. However this trouble is being done away with as comfortable walkways are being made where needed.

Peterson's Leap and the Abyss are deep, rough, ugly looking holes. The Leap is some 30 or 40 feet deep and is found to connect with another route. Across the Abyss is an unexplored region whose mysteries are yet unsolved. There are many lateral branches running in different directions and various distances, sometimes lending back to the Main Route in their windings. There is also the Upper Chamber which is difficult of access, being some 30 or 40 feet above the main body, but which amply repays the tourist.

Pools of clear, cool water are found in various parts of the cave, some running, some without current; those which run frequently and rise after heavy rains but some run down to their accustomed stage. The distance of the largest explored route is about 3 miles.[7]

A deed book says that J.M. Rice, C.P. Warfield, and J.P. Gracey bought Dunbar Cave in Chancery Court in 1882. Several tracts of land are involved and apparently they sold frequently, which is very confusing.

Uncle Josh Rice was manager for the 1883 season and held free concerts at the cave "inside and out" with Dick Jackson vocalist and Sam Aeglen, instrumentalist. The 4th of July Party featured a bran dance.

The August 24, 1883, edition of the Nashville *Daily American* reports a meeting of the Montgomery County Agricultural Association that was attended by "3,000 farmers with their sturdy sons and their fair daughters." The article notes that: "Very early the young people secured the space in the mouth of the cave—a more delighted and picturesque position could hardly be found—and were indulging in the mazes of the dance." The article goes on to describe speeches, exhibits, and a huge barbecue as the Grande Finale.

"In 1884, Colonel A.G. Godlett and H.C. Merritt purchased the hotel and property. They expanded the original hotel to accommodate approximately 200 to 300 guests. Hotel rates were $8.00 per week for board and lodging; special rates were made for families and

---

7   *Ibid.*

Figure 5-9. Billyfrank Morrison in a trench dug during commercialization. Photo by Larry E. Matthews, January 15, 2005.

ly issuing from the mouth of the great cave. While here they looked out upon the beautiful valley below and admired the natural amphitheater formed by the surrounding ridge covered with grand forest trees and encircling the shady, level basin a quarter of a mile around and enlivened by music from a rippling stream that winds its way through the valley from the gushing spring above. This valley served as the Fair Grounds intermittently until 1920. For ten years, not consecutively, the Cave Fair and Stock Show were held at Dunbar's Cave in late August.[9]

The main events of these fairs were conducted in what was called the Amphitheater, which was located where the lake now stands. A Pagoda was erected in the center of this stock ring area for the use of the judges and fair officials. The grandstand was erected on the higher ground to the west and northwest of the amphitheater.[10]

The 1885 *Woodbridge's Directory of Clarksville* still lists Rice as the proprietor of

those spending the season. The hotel management also advertised the availability of feed and livery stables to further accommodate their guests. J.H. Bowling and J.M. Rice were the hotel managers."[8]

On July 5, 1885, C.P. Warfield was host to the executive committee of the Stock Breeders and Farmers Association. He took them on a tour of his farm and then ended up at the cave. They enjoyed the invigorating breeze constant-

---

8   Charles M. Waters, op. cit.

9   Priscilla Weathersby, typed document on file at Dunbar Cave State Natural Area, dated 1981.

10   Paul Hyatt, Marie Riggins, Ralph Winters, and Thurston L. Lee, "One Hundred Years of County Fairs in Montgomery County, Tennessee," August, 1960.

Dunbar Cave

Dunbar Cave. By now, Dunbar Cave is a well-established commercial operation. A lengthy description of the cave tour is given:

The proprietors have so widened and improved the various passages that the tourist can make the trip through with very little fatigue. Among the places of notice are the Counterfeiters Room which was occupied by such a band in 1842, Music Hall, about half an acre in extent and from fifteen to thirty feet high; the Ball Room, fifty by four hundred feet in extent. Around the river, the tourist comes to Rocky Mountain, the Cathedral, Dunbar's Coffin, and Spray Hall. The Cathedral is a solid mass of granite on all sides and on the entrance from below, the sides resemble draped curtains and the ceiling of frescoing, beautiful beyond description. To complete the resemblance, at one end is a pool of pure water in solid rock, answering for a baptism in rock. Spray Hall is also a solid cone, about fifty feet high, from the top of which flows a stream of water, becoming spray is it reaches the bottom.

Then comes the Saltpeter Mine, Lovers Leap, Petersons Leap, Relief Hall, Great Relief Hall, and Independence Hall. The latter embraces about two acres of surface, with hundreds of columns interspersed through and around it in all masses of fantastic shapes. In this hall is Solomons Porch, Jacobs Well, Rebeccas Seat, the Happy Dutch Family, the Elephant, the Monkey, the Petrified Coon, the Allapus (sic.) Wallapus, the Irish Potatoes, etc.

From this Hall lead eight different caverns, the Crystal Palace, by far the loveliest of all. It is three stories, the roof of each profusely studded with white and colored stalactites. On the opposite side is a beautiful formation resembling very large tobacco leaves, each leaf producing a different tone, distinct and clear, which, under the touch of a skillful manipulator, could be made to produce the sweetest music. The Ice Pond had the appearance of a pool of clearest water half frozen over. The Diamond Grotto derives its name from myriads of small stalactites and its sparkling columns. By a circuit of Petersons

Figure 5-10. "INDEPENDENCE HALL" and "O. L. S."
neatly written on the ceiling in Independence Hall.
Photo by Larry E. Matthews, January 15, 2005.

*Figure 5-11. "Marye Tyler. – Walter Scott. – Aug 26, 1888" written
in black paint on the cave wall in Indian Trails Passage.
Photo by Larry E. Matthews, January 15, 2005.*

Leap, the effect of a most realistic sunrise is produced.

The mouth of this cave is fifteen feet wide and eight high, from which is emitted a steady, stiff breeze that renders the atmosphere uncomfortably cool in the hottest weather.

The *Goodspeed's History of Tennessee*, which was originally published in 1886, also describes Dunbar Cave in great detail. That description is as follows:

Dunbar's Cave is situated about three miles from Clarksville on the Russellville Pike. Though inferior in size to the Mammoth Cave in Kentucky it is second in beauty to none in the country. The mouth

of the cave is 15 feet wide and 8 feet high, and from the cave is emitted a stiff, uniform breeze, from which fact it is known as a "blowing cave." Here a temperature of from 50 to 75 degrees is maintained the summer through. Among the places of note and objects within the cave are the following: "Counterfeiter's Room," so named from having been occupied by a band of counterfeiters in 1842; "Music Hall," which is about half an acre in extent, and from 15 to 30 feet high; the "Ball Room," which is 50 by 400 feet in extent; "Dunbar's Coffin," "Spray Hall;" "Rocky Mountain;" and the "Cathedral;" the "Saltpeter Mine;" "Lovers' Leap;" "Peterson's Leap;" "Relief Hall;" "Great Relief Hall;" and "Independence Hall." In the latter hall, which embraces about two

*Figure 5-12. "Miss Annie Mimms." written in black paint on the cave wall.
Photo by Larry E. Matthews, December 7, 2004.*

acres in extent and contains hundreds of columns interspersed throughout in all kinds of fantastic shapes, are "Solomon's Porch," Jacob's Well," "Rebecca's Seat," the "Happy Dutch Family," the "Elephant," the "Monkey," the "Petrified Coon," the "Irish Potatoes," and the "Willapus Wallapus." Eight different caverns lead from this hall, the "Crystal Palace" being by far the most beautiful. It is in three stories, the roof of each being studded profusely with white and colored stalactites. The "Ice Pond" and the "Diamond Grotto" are named from their similarity to the objects after which they are named.[11]

In 1886 a frame hotel is built near the mouth of the cave. Managers of the hotel were Mr. and Mrs. W.A. Shelby. A livery stable was also erected and W. Bryant Whitfield was the proprietor. This hotel was known as the Cave Hotel and was not the same as the Idaho Springs Hotel.

Also in 1886, a large number of small lots were sold for use as building sites for cabins.

An article in the September 30, 1886, edition of the Nashville *Daily American* gives the following account:

Clarksville, Tenn., Sept. 29.—(Special)—The great sale of building lots at Dunbar's Cave, Saturday, Oct. 2, is attracting lively interest and will be largely attended. The value of the cave air and splendid waters, recently developed as curatives for most all summer diseases and a pleasant retreat from the pressure of extreme heat, has become so generally known. Many prominent citizens have already purchased lots and are building neat, cheap summer cottages for their families. This sale disposes of all the lots that will ever be sold, and will be the means of building up a beautiful summer village of about 200 cottages. This is the cheapest and best summer resort in the South, where the temperature stands the year round at about 76 degrees. Therefore, the sale is of interest and importance to those wanting a pretty, cheap home in the midst of a healthful summer resort. Strangers are expected from Memphis, Louisville, Nash-

---

11 Goodspeed, *op. cit.*

ville, and other points.

Some of these cabins existed until just a few years ago and were finally removed when a new home was built on the old Idaho Springs Hotel property.

The property changed hands again in 1887. According to Deborah Lingle Gillis: "Due to J.M. Rice's death, the cave property was sold to A.B. Barbour of Bardstown, Kentucky. It was sold for $19,815 which would be a very large sum of money for the times."

There was a newspaper notice of a Stock Show and Reunion to be held on August 23, 1887. The article notes that there would be a: "Special train run each way between Cherry Station and Clarksville. The train would stop at Stacker's line within a few hundred yards of

the cave. Warrens C.C. Kent of Evansville will play for the dance."

On June 29, 1888, the Baptist Picnic was held at Dunbar Cave: "They came in two herdics, 8 furniture wagons, and a number of private conveyances. The herdics are new fangled things which are rainproof. Some of the picnicers got out and walked in the rain down the steep hill that leads to the mouth of the cave because they were afraid of the herdics. A herdic was a kind of low-hung cab, usually with two wheels, with side seats and entrance at the back. They picnicked at noon and went into the Cave. Inside red lights burnt at intervals with the other light out only increased the sublimity of the scene.

In 1889 the hotel was under the management of Mr. and Mrs. T.L. Yancey. A string

Figure 5-13. "Geo. Kirby, Kid Trimble, Annie Harrison, Sukie Shelby, Paula Shelby, Lisa Wilson, Wally Shelby -- July 17, 1888" written in black paint on the cave wall. Photo by Larry E. Matthews, December 7, 2004.

band was engaged for the season. The opening ball was held in the mouth of the cave.

1890 was another busy season. Whitford Bates & Company took a contract for a gas plant at the cave which "should be ready for the opening and will furnish sufficient gas to light the hotel, the cave, and the stable." One account notes that: "The young people had an enjoyable time at the opening ball at Dunbar Cave Thursday night. Two bands furnished music."

In June, 1892 the Montgomery County chain gang is reported to be spending the summer at Dunbar cave and "making a good road to the cave." For June and July picnics are "planned and executed by Carney P. Lyle and Thomas E. McReynolds." The report goes on to say: "In June the picnicers went to the mouth of the cave and some went over to (the) Springs, where they found jolly, fat Pete Johnson hard at work. He said he was fixing the ground for croquet and lawn tennis. Others went up to the Hotel to see Mrs. Johnson, who

made it pleasant for them. The July picnicers were Methodist Sunday Schoolers." Manager Thompson had the Bates Orchestra come down from Idaho (Springs) and play several selections.

On October 10, 1892, the Dunbar Cave Hotel was destroyed by fire together with it's contents. According to the report: "The hotel was a frame building erected about three years ago by Mr. Barbour of Kentucky, owner of the cave property, at a cost of about $5,600. This is not the same building as the Idaho Springs Hotel. In fact, Pete Johnson, the proprietor of the Idaho Springs Hotel, allowed Mr. Thompson and his family who were managing the Cave Hotel to move into the Idaho Springs Hotel until they could find another place to live.

In 1893 W.T. Haynes leases Dunbar Cave for the season and gives notice that a guide will be there during the summer. In 1895 O.D. Thompson was the proprietor and John Rice served as cave guide.

Figure 5-14. "Lillian Leftwich – Edgefield, Tenn." written in black paint on the cave wall. Photo by Larry E. Matthews, December 7, 2004.

Figure 5-15. "Herculanaeum" written in white paint on the cave wall. Photo by Billyfrank Morrison.

The Willapus Wallapus

As noted earlier in this chapter, the *Goodspeed's History of Tennessee* mentions a cave formation in Independence Hall that is named the Willapus Wallapus. Extensive research by the author has failed to reveal exactly what a Willapus Wallapus was, but we must conclude that it was some sort of mythological beast. At any rate, the people in the Dunbar Cave area knew what it was and referred to it from time to time. Jesse Glass found such an account that was printed in the *Tobacco Leaf Chronicle* in the February 5, 1892 issue:

A Ferocious Animal. Ingram's Willapus Wallapus in the Ross View Neighborhood. To the *Tobacco Leaf Chronicle.*

Some wild and ferocious animal has been playing havoc in the Rossview neighborhood. It seems to delight in slaying mules and dogs. An old darky living on Red River had two mules killed by it. He had them confined in a small lot and they were both killed one night last week. One of them was half devoured and the other partially.

It visited John Bellamy's the next night and devoured two of his dogs. It seems to confine its diet to mules and dogs, never returning to a second meal to what is left of the carcass, but requiring a fresh victim every night.

George McMurray saw it and while he was trying to remember the different illustrations of wild animals he had seen in natural history, in order to name it, it sprang at the mule he was leading and he left without arriving at any conclusion as to what it was. The last heard of it was at Ase Randle's; it attacked one of his colts and the dam kicked it through a barbed wire fence, where it left a section of its tail and three claws hanging on the barbs of the wire.

The neighborhood has organized a big hunt for it and will meet at McMurray's store next Saturday at 10 o'clock A.M. Everybody is invited.

Charley McCurdy, who has shot tigers in Africa (sic.), killed Grizzleys in the mountains of California, is selected as the captain of the party.

The program is to send the dogs in front and after it has devoured half a dozen dogs we will venture in shooting distance.

(Signed by)
ONE OF THE HUNTERS

Dunbar Cave

Jesse Glass found yet another article in the March 25, 1892, *Tobacco Leaf Chronicle* which is titled: "A Great, Ferocious Animal Killing All Of the Hogs About Rossview." It reads as follows:

A strange and mysterious animal has invaded the Rossview neighborhood, causing a good deal of sensation. Mr. J. F. McMurray, a man who never becomes excited, goes off at tangents, indulges in practical jokes, or deviates from the line of plain truth—a man of an investigative turn of mind, says he cannot imagine what it is that is disturbing the quietude of that vicinity; but after thorough investigation he is satisfied that some strange, ferocious wild beast is roaming around destroying everything it can get its mouth on. It has visited the farm of Col. J.B. Killibrew several times, committing depredation and killing his hogs. It has eaten several large, fat hogs, leaving nothing but the feet and larger bones. One night a sow and eight pigs were devoured at a single meal. It travels over the wheat fields, leaving tracks as large as a man's hand, with long claw prints.

It was tracked out to the river, leaving great patches of hair where it broke through the barbed wire fence. Mr. McMurray, as we understand, has fleeces of this hair which is three to four inches long. The color next to the skin is white, while the outer ends are of reddish or fox color. The beast was tracked to a cave in the river bluff. Just at the mouth of the cave it seems that the animal stopped on a flat rock and picked the cockle burrs out of its hair. The investigators also have a collection of these burrs with the hair sticking to them. No man, however, will venture near that cave alone. The neighbors all go together with butcher knives and double-barrel guns, heavily charged. These men are going nights in squads of four hunting the thing. It must be caught and killed or there will be no hogs left in this vicinity, and after the hogs and cows have been exhausted what will become of farm hands? This is the serious question.

Unfortunately, today, no one seems to be able to tell us exactly what a Willapus Wallapus was or what it looked like.

Chapter 6

Legends, Rumors, and Outright Lies

Upon researching the long and interesting history of Dunbar Cave, it becomes readily apparent that there are some stories regarding this cave that appear to have little, if any, basis in fact. Some of these might best be described as legends, which are stories passed down from person to person and considered by some to be historical facts, but those facts are not verifiable. Less reliable stories could be considered rumors, since it is clear that they cannot be proven historically and they are even less likely to have originated from true stories. And, without a doubt, some of the stories concerning Dunbar Cave are just outright lies. The commercial cave business definitely falls in the category of entertainment. Some cave owners and cave guides have always considered that a tall tale concerning famous people or events will enhance the popularity of their cave. Many of these stories are passed down from one guide to another and, as would be expected, each guide gives a slightly different version. Therefore, these stories change and become greatly embellished over the years. Below are a few of the more interesting legends, rumors, and outright lies concerning Dunbar Cave.

The Counterfeiters Den

A counterfeiting ring making counterfeit coins is said to have operated around Clarksville in 1842. Two Clarksville brothers, who were reputed to be the local ring-leaders of an east Tennessee ring, were seen entering Dunbar Cave. The sheriff formed a posse and entered

the cave only to find that the men had disappeared, evidently through a back entrance. The posse found some of the counterfeit coins, but failed to find the back entrance. (Eleanor S. Williams, 1990) A newspaper article in the Chattanooga *Daily Times* (June 6, 1893, page 5, column 5) gives the following account:

OLD COUNTERFEIT COINS

Found Near Dunbar's Cave—The Work of an Old Time Counterfeiter.

Special Correspondence Chattanooga Times.

Clarksville, Tenn., June 5.—On Saturday while digging for fish worms near Dunbar's Cave, four miles east of here, George Herdman unearthed several pieces of counterfeit coin that were identified by older citizens as the product of the dies that Dick Cherry manipulated away back in the '20s. The coins were nearly destroyed by rust. One half dollar was polished until the date, 1820, could be plainly seen. It was the regulation American coin. Another dollar piece was made after the Mexican fashion. It had an Indian's head on it, surmounted by the rays of the sun. All the coins were made of lead. Cherry plied his trade in a room about thirty or forty feet from the mouth of the cave to the left. The room still shows signs of its early illegal use. There is a flue left, but the dies are all gone. Cherry was arrested, but never tried. Revenue officers attempted to guard him over night in

Figure 6-1. Counterfeiters Den.
Postcard from the Billyfrank Morrison Collection.

the cave, but being well acquainted with all its secret entrances, exits, and grottos, he escaped and was never recaptured.

At that time, the Montgomery County Fairgrounds and Racetrack were near Dunbar Cave and betting on horse races was quite common. There were few silver mines in the United States prior to the discovery of the Comstock Lode in Nevada in 1859. As a result the United States Mint had produced relatively few silver coins by 1842. Therefore, foreign silver coins of many varieties circulated freely. In fact, Spanish and Mexican coins were legal tender until 1857. However, it is doubtful that such a counterfeiting operation as described above could have taken place in the cave without the owner's knowledge. Even more important, the fire necessary to run such an operation could not have been anywhere in the cave, much less such a small room, without overcoming the operators with its smoke.

Another version of this story is that the counterfeiters hid some of their counterfeit coins in Dunbar Cave. This would be more likely. According to one version of the history of the Dunbar Cave property, William Tate sold the cave to Charles Cherry in 1814. When

Cherry passed away he left the property to two of his sons. Both sons were to share the cave and the springs. Now, assuming this is true and that the two sons were counterfeiters, it would make sense that they would hide the counterfeit coins in their own cave. In this version of the history of Dunbar Cave, the oldest son, Thomas Cherry is convicted of holding and passing counterfeit coins in 1839.

The Counterfeiters Room, shown on the commercial tour for many years, is approximately 700 feet from the entrance and on the west side of the main passage. During the commercial period, there were two mannequins in this room to represent the counterfeiters.

The Visit By Jenny Lind

Jenny Lind, a world famous singer known as the *Swedish Nightingale*, performed in Nashville, Tennessee, in 1851. According to the Tennessee Genealogy Web site, while she was in Nashville she was told about Dunbar Cave and she hired a carriage to drive her to the site. They go on to state that: "To test the acoustics inside the cave she sang *The Last Rose of Summer*." The Web site includes a photograph that shows her name written on the ceiling with the soot from a candle and the date 1851. The

Figure 6-2. Egyptian Grotto.
Postcard from the Billyfrank Morrison Collection.

May 21, 1967 *Clarksville Leaf-Chronicle* revisits this story:

> Jenny Lind, the famous "Swedish Nightingale" sang at Dunbar Cave in the Spring of 1851. Of the many great personalities who rode the waves of the Cumberland River and stopped at the old Clarksville wharf, the citizens were sorely disappointed when this famous lady failed to appear when she passed through on March 28, 1851.[1]

Figure 6-3. Photograph of Jenny Lind
by Matthew Brady. Date unknown.

photograph was taken in 1948, nearly 100 years later. During the days of the commercial tours, there was a beautiful doll on display at this point in the cave, supposedly representing Jenny Lind. A photograph of this doll is on file at the office of the Dunbar Cave State Natural Area. (See Figure 6-5.)

My initial research made me believe that this visit to Dunbar Cave never occurred. Miss Lind traveled to Nashville by steamboat, the fastest and best method of transportation in those days. Her boat did dock at Clarksville for 45 minutes, but the people of Clarksville were severely disappointed when she did not even appear on deck. However, the fact that she did not get off the steamboat does not prove that she did not take a carriage to Clarksville a few days later and visit Dunbar Cave.

The *Modern Living* Section of the Sunday,

This agrees with my original research. However, this newspaper article, written by Eileen M. Baggett, describes a new (in 1967) book, titled: *The Lost Letters of Jenny Lind* by W. Porter Ware and Thaddeus C. Lockard, Jr. published in London by Victor Gallancz Lt. After Jenny Lind had become famous in Europe, she toured the United States from 1850 to 1852 under the management of P.T. Barnum. Ap-

---

1   Eileen M. Baggett, "To Test Its Acousitcs 'Swedish Nightingale' Sang In Dunbar Cave," *Clarksville Leaf-Chronicle*, May 21, 1967, pages 1-A, 3-A.

Figure 6-4. Jenny Lind's signature. Photo by Larry E. Matthews, 2005.

parently, after the Nashville concert, Lind and her entourage traveled north to Louisville by land. It is a well known fact that Lind visited world famous Mammoth Cave at this time. According to this article, which references the April 5, 1851 edition of the *Jeffersonian*, Lind also stopped in Clarksville and visited Dunbar Cave. Baggett writes:

> It was at this time she stopped at Dunbar Cave. A party of 50 to 60 Clarksville people accompanied her through the cave. The party discussed the acoustics of the cave and asked her if she would sing. She favored them with "The Last Rose of Summer" and signed her name, in smoke from a lantern, on the wall of the cave. The writing is still discernible.[2]

Unfortunately, there is no quote from Lind's "Lost Letters" to verify that she did visit Dunbar Cave. Perhaps more research will provide proof that she did sing there.

Use During the Civil War

There are rumors that during the early part of the Civil War, the Confederate Army used the cave to conceal horses and troops from the Union Army. Local residents are also supposed to have used the cave to hide their livestock and other items from Union troops. After the area was secured by the Union troops, they are supposed to have used the cave as a hospital for wounded Federal soldiers. It is reported that the cave continued to be used for social affairs during the Civil War, as Colonel S.D. Bruce, commander of the Union troops occupying Clarksville, hosted a picnic there.

Because nearby Fort Henry, on the Tennessee River, and Fort Donelson, on the

---

2   Ibid.

Cumberland River, both fell to Federal troops early in the Civil War (Fort Henry on February 6, 1862, and Fort Donelson on February 16, 1862), there was very little Confederate presence in north-central Tennessee for most of the Civil War. And certainly, no competent leader would hide his troops or horses in a cave where they could be so easily cornered. It is also unlikely that the cave would have been used for a hospital, when there were two existing hospitals in nearby Clarksville. In all probability, the picnic mentioned above was the most exciting thing to happen at Dunbar Cave during the Civil War.

The Ku Klux Klan

After the Civil War, Ku Klux Klan meetings are rumored to have been held at the cave. Deborah Lingle Gillis (June, 1993) mentions a reference that reports: "The Willapus Wallapus and White Sheets were out on July 8, 1902",

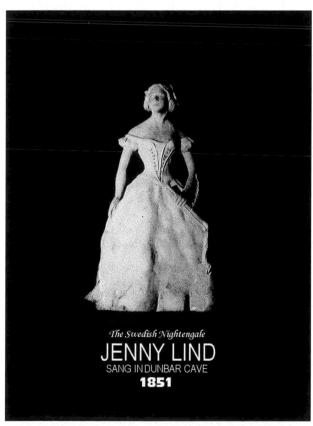

The Swedish Nightengale
JENNY LIND
SANG IN DUNBAR CAVE
1851

Figure 6-5. Photograph of a Jenny Lind doll that was on display inside Dunbar Cave during the commercial tours. Photo courtesy of Dunbar Cave State Natural Area.

Figure 6-6. The Little Volcano. Postcard from the Billyfrank Morrison Collection.

implying the area was still used for Ku Klux Klan meetings.[3] There is no documentation of any Ku Klux Klan meetings at Dunbar Cave.

Unfortunately, no description or definition of the term "Willapus Wallapus" has been found, although it is also mentioned in *Goodspeed's History of Tennessee*. Its association with the Ku Klux Klan is unclear.

Portrait of the Confederate Soldier

There is an interesting drawing of what is supposed to be a Confederate soldier on the east wall of the main passage between the Counterfeiters Room and Concession Rock. Also associated with this portrait is the date 1862 and various assorted writings that are difficult

---

3   This account is from a newspaper article that is included in one of the notebooks on file at the Dunbar Cave State Natural Area Office.

to decipher. Exactly who painted this portrait on the cave wall is unknown, as is the authenticity of the date 1862. However, it is a very interesting portrait showing the clothing and hairstyles of that time period. There is another portrait located in Indian Trails Passage near the Jenny Lind signature. It appears to have some coloration, which may be from an artist's chalks. (See Figure 3-3) Several other smaller portraits and drawing are also found in the Historic Section of the cave.

Figure 6-7. Portrait of the Confederate soldier. Photo by Larry E. Matthews, 2005.

Jesse James Hides Gold in Dunbar Cave

Another persistent rumor was that Jesse James hid gold in Dunbar Cave. Frank and Jesse James are famous outlaws who roamed throughout the United States from the end of the Civil War until Jesse's death in 1882. The James Gang was notorious for robbing banks and robbing trains of their gold shipments. In their first robbery, they stole $60,000 in gold from the Liberty, Missouri bank. Their travels led them through Tennessee where, of course, they robbed several banks. For whatever reason, there were persistent rumors that Jesse James and his gang hid some of this gold somewhere deep inside Dunbar Cave. However, there is absolutely no evidence that they ever visited Dunbar Cave. Interestingly, Jesse James lived in nearby Nashville, Tennessee, under the assumed name of Thomas Howard for several years.

The Back Entrance

Like most well-known caves, Dunbar Cave was rumored to have a "secret" entrance or a "back" entrance. This entrance was supposedly known to the band of counterfeiters and to Jesse James, among others. Naturally, they used this as an escape route to foil the law enforcement officers who were chasing them. And, as in most such stories, no one but the bandits was able to locate and use this "secret" entrance. As real cavers know, it doesn't take many trips through a cave passage to leave a trail, so if there had been such an entrance, it really would not have been that hard to find. But no such entrance exists in the Historic Section of Dunbar Cave.

Chapter 7

Idaho Springs Resort: 1900 - 1930

The manager of Dunbar Cave for 1900 was a Mr Heathman. He reported a big Fourth of July celebration. "William Kleeman prepared a delightful barbeque."

Management changed to Sterling Fort for the 1901 season. He continued to be the manager for the 1902 season.

On June 15, 1903, Cecelia Barbour sold Dunbar Cave to the Electric Street Railway Company of Clarksville. Electric Street cars were also known as trolleys, and were a major form of public transportation before the widespread use of the automobile. They ran on railroad tracks. It was a common practice for street car companies to have a public park at the end of their line to encourage people to ride the street car to that destination, especially on weekends. The Montgomery County Fair for 1903 was held at Dunbar Cave.

W.W. Tate was the manager for the 1905 season. The 1907 Montgomery County Fair featured Madam Lavena, a balloon race, and a parachute drop.

The 1911 Cave Fair and Stock Show was held on August 24–26. There was a Fiddlers Contest on September 1.

On October 15, 1913, the Electric Street Railway Company of Clarksville sold eight tracts of Dunbar Cave property to the Clarksville-Dunbar Cave Railway Company and they sold out to Louis Fisher of Saint Louis, Missouri.

In 1914 Dunbar Cave was purchased by Wesley Drane and Austin Peay. Austin Peay would later become governor of Tennessee (1923–1927) and nearby Southwestern Presbyterian University would be re-named Austin Peay University in his honor.

A 1915 brochure, titled *"A Summer Vacation at Idaho"*, announced that the Idaho Springs Hotel was opening a new and up-to-date Summer Resort *"To the Health-and Pleasure-Seeking Public"* on June 1, 1915. Information in this brochure includes:

IDAHO SPRINGS HOTEL

We open our new and up-to-date Summer Resort

To the Health-and Pleasure-Seeking Public

June 1, 1915

Location

Idaho Springs is situated three miles east of Clarksville, a middle Tennessee town of ten thousand inhabitants, and one and one half miles south of St. Bethlehem, a village on the Louisville & Nashville Railroad, to which special rates are made as far south as Memphis. Conveyances may be had at a nominal cost from St. Bethlehem, and upon notification, will be in waiting.

Figure 7-1. Idaho Springs Hotel. Photo courtesy of Dunbar Cave State Natural Area.

Hotel

The hotel is situated on a good elevation, making it free from dampness and malaria, and consists of three separate buildings, which are so constructed that every room has two outside exposures, thus insuring plenty of fresh air. The dining room, which is also used as a dance hall, is detached from the main sleeping department, thus making it quiet for those who want to sleep. These buildings are new throughout and newly furnished. They are equipped with sanitary toilets and baths, and lighted with acetylene gas. The verandas contain more than ten thousand square feet of floor space, which makes it pleasant for those who do not care to stroll about the grounds.

Grounds

A lawn of about twenty-five acres contains a beautiful hill, sloping to the west, and studded with every variety of forest trees indigenous to this climate. Rustic seats and hammocks are scattered here and there, inviting the weary guest to loiter and rest. At the foot of this hill is a valley, densely shaded by symmetrical sycamores. Through this valley there flows a limpid stream of cold water. This has been harnessed to an under-shot water-wheel, for pumping water to a large tank, whence it is piped throughout the buildings, to the baths, toilets, etc. Near the water-wheel is the play ground, where croquet, tennis, and other out-door sports may be indulged in.

Waters

About one hundred yards down the valley, the wells are located, consisting of White Sulphur, Red Sulphur, and Chalybeate, all within a few yards of each other. The White Sulphur is used mostly for stomach and liver complaints. It has also made wonderful cures of rheuma-

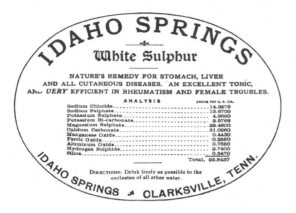

Figure 7-2. Label from a bottle of "Idaho Springs White Sulphur Water." Photo courtesy of Dunbar Cave State Natural Area.

tism, in conjunction with the hot baths. The Red Sulphur is a kidney and bladder water, possessing some qualities that dissolve gall stone. It also affords relief for Bright's Disease, in some cases. The Chalybeate is a good tonic and appetizer, especially when used in conjunction with the other waters.

Fishing

Those to whom an out-door life appeals, can find plenty of recreation and sport in fishing, boating, and bathing at Red River, about five hundred yards distant. This river has been stocked with the finest trout, which are now plentiful. The Hotel owns several boats which are free to the guests.

Other Things We Have

We have the coolest dining room in the State, and the windows are all screened; in fact every door and window in the Hotel is screened. We have country cooking, old style, and plenty of fried chicken. We have plenty of blankets, and you will use them every night during the summer. We have dancing, billiards, cards, croquet, etc. and plenty of quiet places for those who wish to read or sleep. We have as clean a place as there is in the State, excepting none. We have courteous and efficient servants who are glad to do your bidding.

Figure 7-3. Car at the entrance to the Dunbar Cave grounds. "Admission 20 cents, Tax Included." Photo courtesy of Dunbar Cave State Natural Area.

Figure 7-4. The Concession Stand at the Dunbar Cave Entrance.
Photo coutesty of Dunbar Cave State Natural Area.

Figure 7-5. The Concession Stand at the Dunbar Cave Entrance.
Photo courtesy of Dunbar Cave State Natural Area.

Things We Do Not Have

We have no mosquitos. We have no malaria.

Dunbar's Cave

Adjoining the Hotel grounds is the famous Dunbar's Cave. This is said to be the most beautiful cave in the South, both interior and exterior. It is second in size to Mammoth. It is a sure cure for infants with summer complaint. The thermometer stands 56° F, the year round, making it a pleasant refuge for those suffering from heat prostration.

Rates

Rates for June will be $10.00 per week and up, according to the location and number of occupants of room, or $2.50 per day. Children vary from one-half to full price. Colored nurses, one-half price. Special rates to families, and to those spending the entire season. For further particulars address,

J. H. Tate
Rural Route No. 3
Clarksville, Tenn.

In 1915, the Baptist picnic was held on July 20, the Methodist picnic was held on July 22, and the County Fair was in August.

In 1916 Professor McQueen from Southwestern Presbyterian University built a concrete dance floor in the mouth of Dunbar Cave. W.I. Harnes was operating the cave while Mr. and Mrs. J.H. Tate were running Idaho Springs.

W.I. Haynes was the manager for the 1920 season. The opening dance was held on June 12. Dances were held every Tuesday and Saturday night. The Montgomery County Fair was held on August 26–28.

From 1920 to 1923 livestock conventions were held at the resort.

The Presbyterians had their annual picnic at Dunbar Cave on July 7, 1921.

For the 1923 season, dances were held every Tuesday and Friday night. McCoy Miller of Lynchburg, Virginia, performed for the entire season. His band was composed of Rudolph Shinnick playing two saxophones and a clarinet, Dick Winn playing the banjo, Henry Gupton playing the trumpet and trellophone, and Irvin Brunty playing traps (drums).

W.I. Haynes was the manager for the 1924 season. The Fourth of July party featured a noon barbeque and Johnnie Ely and His Melody Boys furnished music for dancing.

In 1931 a group of Clarksville businessmen (John T. Cunningham, H.C. Merritt, Adolph Hach, J.T. Cunningham Jr, W.E. Beach, C.C. Beach, and B.F. Runyon) bought the property and formed the Dunbar Cave and Idaho Springs Corporation, to redevelop the resort as "The Showplace of the South."[1] According to Jim Whidby:

"Subsequently, an earthen dam was built across the valley, near the cave entrance to form a 15-acre lake fed by the stream issuing from the cave. The lake was stocked with fish and boats were made available for rental. A bridge was constructed across the lake near the cave entrance for access to nature hikes on the opposite side. A modern concrete swimming pool with three diving boards and a bathhouse were built replacing a smaller pool, which had existed in the area of the new lake. In addition, tennis courts and other recreational facilities and restrooms were also constructed."[2]

An undated ad for Dunbar Cave gives the following description of the facilities:

Wonderful, new modern, concrete, regulation swimming pool, size 72 X 120 feet. Every sanitary precaution; continuous flow of fresh, cold water. Qualified life guards and swimming instructors. Beau-

---

1 Jim Whidby, "Idaho Springs and Dunbar Cave Resort—The Showplace of the South," *Journal of the Cumberland Spelean Association*, vol 6, no. 1, pp 7-12.
2 *Ibid*.

Figure 7-6. Looking out the Dunbar Cave Entrance on a summer day.
Colorized postcard from the Billyfrank Morrison Collection.

tiful, commodious, well appointed bath house. Large sand beach, lounging chairs and umbrellas. Water carnivals, fancy diving and swimming contests.

See "the most beautiful cave entrance in America"--always cool the hottest weather—a refuge from the heat.

Facilities for picnics and bridge parties—Cool, shaded pavilion—Lunches obtainable on the grounds.

Beautiful dance floor in mouth of cave, overlooking lake—always cool the hottest weather, 56° year 'round,--makes summer time dancing a real pleasure.

Dancing Tuesday and Saturday nights—other nights on occasion.—Finest orchestras in the South.

Beautiful 15-acre artificial lake, well supplied with large- and small-mouth *bass, blue gills, crappie,* and *sun-fish.*—brooding ponds insure restocking—*boats for fishing* or *pleasure.*

The large bath house for the swimming pool was built in 1933. Today, this building is the Office for the Dunbar Cave State Natural Area. There were also tennis courts, a bowling alley, a baseball diamond, and bridle paths. Electric lights had been installed in the cave by this time.

On June 11, 1934, there was a celebration of Clarksville's Sesqui-Centennial at Dunbar Cave. That night there was a reception with Governor and Mrs Hill McAlister, Dorothy Dix, Congressman, Mrs Joseph L. Byrns, and local dignitaries, all in the receiving line. There was a large fireworks display at 10 o'clock and dancing to Francis Craig and his Orchestra.

Famous bands that performed at Dunbar Cave in the 1930s and early 1940s included Benny Goodman, Tommy Dorsey, Lena Horne, Kay Kyser, Ozzie Nelson, Glen Miller, and Guy Lombardo. These will be covered in more detail in Chapter 8 – The Big Band Era.

Why Dunbar Cave?

At this point, it seems appropriate to stop and ask: Why Dunbar Cave? Why did Dunbar

A PART OF THE LAWN — PHOTO BY FLYNT

IDAHO SPRINGS HOTEL

We open our new and up-to-date Summer Resort

To the Health- and Pleasure-Seeking Public

June 1, 1915

A CORNER IN THE DINING ROOM — PHOTO BY FLYNT

Location

IDAHO SPRINGS is situated three miles east of Clarksville, a middle Tennssee town of ten thousand inhabitants; and one and one half miles south of St. Bethlehem, a village on the Louisville & Nashville Railroad, to which, special rates are made as far south as Memphis. Conveyances may be had at a nominal cost from St. Bethlehem, and upon notification, will be in waiting.

Hotel

THE HOTEL is situated on a good elevation, making it free from dampness and malaria, and consists of three separate buildings, which are so constructed that every room has two outside exposures, thus insuring plenty of fresh air. The dining room, which is also used as a dance hall, is detached from the main sleeping department, thus making it quiet for those who want to sleep. These buildings are new throughout and newly furnished. They are equipped with sanitary toilets and baths, and lighted with acetylene gas. The verandas contain more than ten thousand square feet of floor space, which makes it pleasant for those who do not care to stroll about the grounds.

Grounds

A LAWN of about twenty-five acres contains a beautiful hill, sloping to the west, and studded with every variety of forest trees indigenous to this climate. Rustic seats and hammocks are scattered here and there, inviting the weary guest to loiter and rest. At the foot of this hill is a valley, densely shaded by symetrical sycamores.

MOUTH OF DUNBAR'S CAVE — PHOTO BY PROFIT

Figure 7-7. Brochure titled "A Summer Vacation At Idaho." Dated June 1, 1915. (continued on the next page) Courtesy of Dunbar Cave State Natural Area.

Dunbar Cave

INDEPENDENCE HALL—Dunbar's Cave   PHOTO BY PROFIT

croquet, etc., and plenty of quiet places for those who wish to read or sleep. We have as clean a place as there is in the State, excepting none. We have courteous and efficient servants who are glad to do your bidding.

Things We Do Not Have

We have no mosquitos. We have no malaria.

Dunbar's Cave

ADJOINING the Hotel grounds is the famous Dunbar's Cave. This is said to be the most beautiful cave in the South, both interior and exterior. It is second in size to Mammoth. It is a sure cure for infants with summer complaint. The thermometer

Cave become the social focal point for Montgomery County? There seem to be a variety of factors that account for this, and any one of these factors alone would not explain the phenomenon.

Perhaps as important as anything was the **natural air conditioning**. The large entrance to Dunbar Cave blows out large amounts of cold air during hot weather. Since artificial air conditioning was unknown in the 1800s, this was a rare commodity indeed. But equally important was the **location**. Dunbar Cave was located within sight of a major road in the days when good roads were few and far between. Therefore, Dunbar Cave was accessible to residents from all over Montgomery County and even the surrounding counties. However, once people got there, the next thing that they needed to accommodate large crowds and exhibits was a **large, flat field**. This is one of the features that attracted Isaac Rowland Peterson and Thomas Dunbar: large, flat fields of fertile soil that were ideal for farming. These same large fields were also ideal for large gatherings. Another necessity for any large crowd is a reliable source of **drinking water**. This requirement was met by the constant flow of cool water from the spring at the mouth of Dunbar Cave. Equally important, the Idaho Springs provided at least three different types of mineral water, which were highly prized in the 1800s. And

last, but not least, Dunbar Cave itself provided **entertainment** in the form of guided tours. Dunbar Cave was relatively easy to develop as a commercial cave, because the Historic Section is generally level, with a dirt floor and very little breakdown. The cave was more interesting than most, since it offered a wide variety of features, including a large underground stream, walking-size passages, large rooms, and beautiful formations. Dunbar Cave made an ideal tourist cave. Clearly all of these features combined made Dunbar Cave the most ideal place for social gatherings in Montgomery County.

Starting with all the above natural attributes, the various owners of the cave were shrewd enough to promote the cave widely and to host a variety of meetings there. Equally important, they provided other forms of entertainment, especially music and dancing. Food was clearly available and no doubt was a major source of income for the owners. By establishing a hotel and cabins, daily use of the facilities was assured. Once Dunbar Cave was established as the most important social point in Montgomery County, it became self-perpetuating for nearly a century. Only the advent of modern times caused the demise of Dunbar Cave as the focal point of Montgomery County entertainment. The causes of this will be discussed in Chapter 10 – The McKay King Era.

Chapter 8

The Big Band Era: 1931 – 1947

According to the *Tennessee Encyclopedia of History and Culture*, in 1931 and 1932 a group of Clarksville businessmen acquired the old Idaho Springs and Dunbar Cave property and expanded the recreational facilities, especially at the renovated and enlarged hotel, which now fronted an improved federal highway. The investors built a new dam that increased the lake size to approximately 15 acres. The new complex also included a bathhouse, tennis courts, cabins, and a modern concrete swimming pool. Throughout the Depression, Dunbar Cave was a popular resort and hosted big band concerts, including such famous names as Benny Goodman and Tommy Dorsey. Cave historian Jim Whidby reports that:

> At the cave entrance (35 feet wide and 10 feet high), the old dance floor was enlarged and a concession stand was built along the cliff wall with a railed, concrete-floored terrace atop the concession stand, which overlooked the dance floor. A lower deck supported a fine arcade. There was a bandstand and some of the nation's finest "Big Bands" performed there. Names such as Kay Kyser, Guy Lombardo, Artie Shaw, Shep Fields and his Rippling Rhythms, Blue Baron, Tommy Dorsey, Glenn Miller, and Benny Goodman are but a few of the famous entertainers who performed inside the mouth of Dunbar Cave. These bands would often be "between dates" on their Louisville to Memphis circuit. Many of these performances would be insured for $2,000 with Lloyds of London against cancellation due to bad weather.[1]

An electrical wiring system was installed in the cave and tours were conducted on a regular basis. During this time the Idaho Springs Hotel underwent remodeling, and Henry Merritt, one of the stockholders, became manager of the hotel. Leonard Sowell was the desk clerk and Mrs L.L. Dubois managed the dining room. The mineral waters of Idaho Springs were available to guests, along with excellent food. Each mineral spring had a "gazebo" covering and a concrete floor with a hand pump installed to send the bubbling mineral water to the top. A bowling alley was also added to a long list of recreational amenities.

The May 28, 1933 *Clarksville Leaf-Chronicle* announced that Dunbar Cave would open for the season with a banquet at the Idaho Springs Hotel to be followed by a dance.

The Big Band Era

Dunbar Cave was host to some of the most popular bands in America during this time period. Here is a list of some of those bands, arranged by the year that they played at the cave:

---

1 Jim Whidby, "Idaho Springs and Dunbar Cave Resort—The Showplace of the South," *Journal of the Cumberland Spelean Association*, vol 6, no. 1, pp 7–12.

1933	Francis Craig[2]	1934	Lee Cannon
			Francis Craig
			Bernie Cummins Band
			Jimmy Gallagher
			Phil Holtz
			Isaac Jones
			King Oliver

2   My mother, to whom this book is dedicated, was a huge fan of Francis Craig. Craig (1900–1966) was a pianist with a dance band who lived in Nashville, Tennessee. His most popular song was *Near You*, which was the most popular song in the United States for 17 consecutive weeks in 1947.

Figure 8-1. Idaho Springs Hotel.
Postcard from the Billyfrank Morrison Collection.

Figure 8-2. View of lake and pavilion at Roy Acuff's Dunbar Cave.
Postcard from the Billyfrank Morrison Collection.

Figure 8-3. Independence Hall.
Postcard from the Larry E. Matthews Collection.

Figure 8-4. Diamond Grotto.
Postcard from the Billyfrank Morrison Collection.

Beasley Smith and his WSM
    Orchestra

1935   Agnes Ayers
    Petro Brescia
    Henry Busse
    Francis Craig
    Johnny Endicot
    Jimmy Gallagher

Tom Gray
Fats Kelley
Phil Lorner
Vincent Lopez
Jimmy Mansfield
Freddie Martin
Harry Sosnik
The Fourteen Virginians

1936 Gus Arnheim
   Esther's Blue Ribbon Band
   Ace Brigade and His Band
   The Southern Colonels
   Francis Craig
   The Chicago Follies
   Red & Gray
   Horace Halley
   Johnny Hamp
   Tiny Hill
   Kay Kyser, the Dean of the
    Kollege of Musical
    Knowledge
   King Oliver
   Jimmy Osborne
   Red Pepper
   Marjorie Rainey
   Pop Rinehart
   Noble Sissle and his Orchestra
    (with Lena Horne)
   Jack Stalcup
   Bobby Tyldesly

1937 Gus Arnheim
   Dick Cisne
   The Southern Colonels
   Jackie Coogan and His
    Orchestra

Figure 8-5. The Bridal Veil. Postcard from the Billyfrank Morrison Collection.

Figure 8-6. Old postcard showing the Arches below the cave entrance. Note bridge across lake in the foreground. Postcard from the Billyfrank Morrison Collection.

Figure 8-7. A typical tour group in Dunbar Cave, about the 1930s.
Photo courtesy of Dunbar Cave State Natural Area.

Francis Craig
Art Kassel and His Castles in
    The Air
Little Jack Little and His
    Orchestra
Freddy Martin
Adrian McDowell
Red McEwen
Carl "Deacon" Moore

1938   Charlie Agnew Band
    Dick Cisne Band
    Johnny Hamp Band
    Woody Herman Band
    Isham Jones and His Band
    Henry King Orchestra
    Phil Levant

Red McEwan
Jimmie Reed
Joe Sanders and His Band
Artie Shaw (with Billie
    Holliday)
Jack Stalcup

1939   Charlic Agnew Band
    Blue Barron
    Ace Brigade
    Henry Busse
    Francis Craig
    Bernie Cummins
    Happy Felton
    Jan Garber
    Ray Johnson
    Adrian McDowell

Red McEwan
Jack Stalcup
Nick Stewart Band

1940   Bill Bardo
Blue Barron
Del Courtney Orchestra
Francis Craig
Ella Fitzgerald and Her Band
Ted Florita
Johnny Hamp
Art Kassel and His Orchestra
Buddy Kay
Freddy Martin
Red McEwan
Ozzie Nelson and His Orchestra
    (with Harriet Hilliard)
George Olsen
Ted Fio Rito
Jan Savit

1941   Blue Barron
Lou Breeze
Del Courtney
Jan Garber
Jack Gregory
Woody Herman Band
Isham Jones
Herbie Kaye
Pinkey Tomlin Band

1942   Blue Barron
Bailey Burton
Francis Craig
Jan Garber
Horace Holley
Ina Ray Hutton and Her
    Playboys
Clyde Lucas Band
Charlie Nadgy's Band
Will Osborne and His
    Orchestra

Figure 8-8. Photo of crowd seated at tables on the dance floor and on the roof of the concession stand. Photo courtesy of Dunbar Cave State Natural Area.

Figure 8-9. Couples dancing on the dance floor in the cave entrance.
Photo courtesy of Dunbar Cave State Natural Area.

Jack Stalcup and His Band

1943   Closed

1944   Francis Craig
        Barney Rapp and His Orchestra

Other Big Bands and performers that are reported to have played at Dunbar Cave, for which the year is not known, include:

Count Basie
Owen Bradley
Rod Brasfield and his Blue
    Seal Pals
Duke of Paducah
Ferlin Huskey
Johnny Miller
Don Roy and his Five Comets
Faron Young

Dunbar Cave opened for the 1939 season on May 18, with W.E. Beach as manager. Beach announced that the Hotel was being repainted and extra bathrooms were being added. The opening dance on June 14 featured Francis Craig and His Band. On June 27 at 8:00 P.M. there was a Professional Square Dance Exhibition, followed on Thursday by a dance with Ace Brigade, and another dance on Saturday night with Red McEwan. Saturday night also featured a fireworks display.

On December 7, 1941, the United States was attacked by the Empire of Japan at Pearl Harbor. After the attack, the United States entered World War II. Dunbar Cave remained open. A huge military base, Fort Campbell, was built nearby and soldiers would soon be visiting the cave for entertainment.

The 1942 season opened on Friday, May 15, with W.W. Dunn as manager. The opening dance, as usual, featured Francis Craig and His Band. Admission was $1.65 at the gate.

On May 22, 1943, it was announced that

Figure 8-10. Singer Dottie Dillerd performs in the cave entrance.
Photo courtesy of Dunbar Cave State Natural Area.

Figure 8-11. Dancing in the cave was popular in the days before air conditioning.
Photo courtesy of the Dunbar State Natural Area.

Dunbar Cave and the Idaho Springs Hotel would be closed to the public until further notice, in order to arrange early compliance with requirements by the Army authorities. The cave and the hotel opened on June 12, but the swimming pool remained closed. In order to meet Army requirements for the swimming pool water quality, they laid 1,000 feet of pipe to carry water from a spring to the swimming pool. Previously, lake water had been used to fill the pool. Finally, on July 6, the swimming pool opened.

On June 7, 1944, the opening dance featured Francis Craig and His Band. On June 20, Barney Rapp and His 13-Piece Orchestra performed. Due to military curfews, all dances this year began at 8:00 o'clock and concluded at midnight.

Dunbar Cave opened for the season on May 22 in 1945 and Francis Craig and His Band played for the opening dance.

The cave and the hotel flourished during the 1930s and the early 1940s, and, in 1945, William Kleeman, Mayor of Clarksville, purchased the 400-acre resort and W.W. Dunn continued as manager for the next two years.

* * * * * * * * *

In 1948, country music star Roy Acuff purchased Dunbar Cave and the surrounding property and continued the weekend dances in the cave entrance featuring well-known bands. Later, a live country music radio show was broadcast from the site. According to the April 27, 1948, Nashville *Banner*, "The WSM Grand

Ole Opry headliner, who completed negotiations for purchase of the property for $150,000 yesterday afternoon, said that he planned to construct a pavilion with a capacity of 5,000 persons for the presentation of hillbilly music and other entertainment. Acuff purchased the 200-acre tract—including Dunbar Cave, Idaho Springs, a swimming pool, lake well stocked with fish, and two hotels—from Clarksville Mayor William Kleeman." The August 11, 1952, issue of *Newsweek Magazine* states that: "At the Cave, Roy and his immediate family (his wife, Mildred, and a 9-year-old son, Roy Neill) do not stay at the hotel, but live in a big lodge on the edge of the lake.[3]

In the next chapter, we will cover the Roy Acuff era at Dunbar Cave, which extended from 1948 to 1963.

Suggested Reading

P.J. Broome wrote a wonderful newspaper article titled: *A Decade of Dancing at Dunbar Cave*. It was published in the "Cumberland Lore" section of the Leaf-Chronicle in the June, 1993, issue and the July, 1993, issue.

Joseph C. Douglas wrote an essay titled: *Dancing in the Cool of a Cave: Historic Social Use of the American Underground*. This well-documented essay provides historical information on the use of caves for dancing in the United States in historical times. It was published in the January–June, 2007, issue of *The Journal of Spelean History* (Volume 41, No. 1, Issue 131, pp 27-34).

---

3   Jim Whidby, *op. cit.*

Chapter 9

The Roy Acuff Era: 1948 – 1966

Roy Acuff was born in Maynardville, Tennessee, on September 15, 1903, to a prominent Union County family. Roy's father was an accomplished fiddle player and his mother was a piano player. In the days before Radio and TV, people in rural areas often provided their own music and entertainment. So, it is not surprising that Roy also learned to play several musical instruments while he was still a child. By the time Roy was in high school, the family had moved to Knoxville, Tennessee, where Roy showed talent in many fields, including singing, acting, and athletics. In 1929 Roy joined the Knoxville Smokies baseball team, a minor league team for the New York Giants. Unfortunately, Roy suffered a sunstroke during spring training and was seriously ill for several years. In order to pass the time, he practiced playing the fiddle, and decided to pursue a career in music.

Roy's first professional job as a musician came in 1932 when he was hired to work in Dr Hauers Medicine Show. Acuff and the other musicians would perform and once a crowd had gathered, Dr Hauer would sell them his patent medicines. Early patent medicines were usually a combination of alcohol, opium, and occasionally cocaine. People really did feel better, much better ... and fast!

By 1934 Roy Acuff had his own band, which was known as the Tennessee Crackerjacks. They were good enough to perform on the local Knoxville radio stations. Soon, the band changed its name to the Crazy Ten-

nesseans. By 1938 Acuff's band had moved to Nashville, Tennessee, to audition for the Grand Old Opry, where they were able to obtain a contract. At this time they changed the band's name to the Smoky Mountain Boys, a name that they kept for the rest of their careers.

Clearly, Acuff was a shrewd businessman as well as a talented musician. In 1942 he started a business with songwriter Fred Rose known as Acuff–Rose Music. This business became the most important music publishing company in country music and made Acuff quite wealthy. Among the famous people that were signed by Acuff–Rose were Hank Williams and Patti Page. In addition, Roy Acuff himself had many "hit" songs, including *The Wabash Cannonball* (1936), *The Tennessee Waltz* (1948), *The Great Speckled Bird (1955)*, and *I Saw the Light* (1971). It was the money earned in this business and musical career that allowed Roy Acuff to purchase Dunbar Cave in 1948. The Nashville *Banner* announced Acuff's purchase of Dunbar Cave in their April 27, 1948, edition:

The WSM Grand Ole Opry headliner, who completed negotiations for purchase of the property for $150,000 yesterday afternoon, said that he planned to construct a pavilion with a capacity of 5,000 persons for the presentation of hillbilly music and other entertainment. Acuff purchased the 200-acre tract—in-

Figure 9-1. Roy Acuff and the Crazy Tennesseans, about 1936.
Photo courtesy of Billyfrank Morrison.

Figure 9-2. Roy Acuff (seated) and other entertainers.
Photo courtesy of Dunbar Cave State Natural Area.

Figure 9-3. Entertainers perform outdoors to a large crowd at the Amphitheater.
Photo courtesy of Dunbar Cave State Natural Area.

Figure 9-4. A crowded day at the swimming pool.
Photo courtesy of Dunbar Cave State Natural Area.

Figure 9-5. Roy Acuff and the Smoky Mountain Boys. Photo courtesy of Billyfrank Morrison.

cluding Dunbar Cave, Idaho Springs, a swimming pool, lake well stocked with fish, and two hotels—from Clarksville Mayor William Kleeman."[1]

Since Roy Acuff had grown up in rural east Tennessee, Dunbar Cave also served as a retreat from urban Nashville, Tennessee, and the busy business world. The August 11, 1952, issue of *Newsweek Magazine* states that: "At the Cave, Roy and his immediate family (his wife, Mildred, and 9-year-old son, Roy Neill) do not stay at the hotel, but live in a big lodge on the edge of the lake."[2] This lodge was located on the opposite side of the lake from the swimming pool, bathhouse, and ticket office. This allowed more privacy for Acuff. Later, a houseboat was added to the lake, and Acuff and his guests could relax and fish from the boat.

Acuff was a musician, and no doubt he was attracted to Dunbar Cave as the most successfully music venue anywhere in middle Tennessee. Acuff continued the weekend dances in the cave entrances featuring well-known bands, then later added live country music radio shows broadcast from the site. The schedule was to hold square dances on Tuesday and Friday nights and round dances (big bands) on Saturday night. Grand Ole Opry musicians performed at the cave on Sundays. In 1948 it cost 30 cents to attend a square dance and $1.20 to attend a popular dance. Roy continued the tradition of a big Fourth of July party at Dunbar Cave. In 1948 he performed with His Smoky Mountain Boys and Slim Wooten performed with His Colorado Mountain Boys. There was always a large fireworks display and lots of other entertainment, including magicians, jugglers, acrobats, and a talent show.

W.W. Dunn returned as manager. Roy Acuff built an arena and an 18-hole golf course just east of the lake. The resort was subsequently renamed Roy Acuff's Dunbar

---

1   Jim Whidby, "Idaho Springs and Dunbar Cave Resort—The Showplace of the South," *Journal of the Cumberland Spelean Association*, vol 6, no. 1, pp 7–12.
2   *Ibid.*

Cave. Roy's lodge was conveniently located next to the golf course.

Roy Acuff ran for governor of Tennessee in 1948. Tennessee was a strongly Democratic state at the time, so his campaign on the Republican ticket probably never had much chance of success. In fact, the story goes that the Republicans couldn't find a real politician who would run, so they put Roy Acuff on the ticket just to have a candidate. The election was held on November 2, 1948, and Roy did not win the governor's seat. He probably never expected to.

An undated brochure from this time period proclaims the wonders to be found at Dunbar Cave Resort:

DUNBAR CAVE and RECREATION PARK
A PLACE TO SPEND

YOUR CAREFREE HOURS
ALWAYS 56° COOL

Competent guides will show you the wonders that nature has wrought deep in the earth. Everywhere are fantastic formations of stalactites and stalagmites, thousands of years in the making. An underground river disappears and reappears at different points of interest on the tour.

The river holds subterranean animal life, entirely white and without any vestige of eyes. The great antiquity of the cave holds a fascinating history. Known by the Indians into the dimness of time, it has seen savage massacre, been the "vault" of counterfeiters, and has been the site where

Figure 9-6. Paddle boats and boat dock on the lake. Photo courtsy of Dunbar Cave State Natural Area.

Figure 9-7. Paddle boats on the lake and people walking towards the cave entrance. Photo courtesy of Dunbar Cave State Natural Area.

Figure 9-8. Tourists in Independence Hall. Photo courtesy of Dunbar Cave State Natural Area.

Jenny Lind sang on a bridge over the River Styx. She wrote her name nearby with a pine torch; the writing is quite legible even today.

THE MOST BEAUTIFUL CAVE ENTRANCE IN THE WORLD

You will find in Dunbar Cave recreational park the incomparable charm and pristine beauty of one of the nation's most beautiful natural playgrounds.

Have your picnic or play cards in the delicious coolness at the mouth of the cave. Tables are strategically placed for your convenience.

Enjoy the folk music shows every Sunday afternoon and night. Be sure to be here on the 4th of July for special entertainment and fireworks.

Swim in the large, modern pool. Every precaution is taken including continuous flow of fresh, cool water; qualified lifeguards; and swimming instructors. Loll on the sand beach and absorb health-giving rays of the sun.

Popular pedal boats for two are fun. Travel by aquatic foot power while you explore the shoreline.

Regulation boats are available on the 15-acre lake. The Isaac Waltons will find ample reward for their efforts from this well stocked sanctuary.

Stroll down the wilderness walkways,

Figure 9-9. Tourists on the trail along the River Styx. Photo courtesy of Dunbar Cave State Natural Area.

Figure 9-10. Cave Guides tell tourists about the cave.
Photo courtesy of Dunbar Cave State Natural Area.

some of them leading to hideaway havens in the woods. Rustic furniture is there for your convenience. Rest in the murmuring quiet of wildlife's habitat.

Dance under the stars on the large floor near the cave entrance. Soft, sweet strains of a famous orchestra fill every romantic moment.

Don't miss the bird reservation. See the strutting peacock, shy pheasants, rare and beautiful pigeons, swans, ducks, and geese.

Everyone loves the mischievous monkeys. They're unpredictable fun lovers, and completely fascinating to young and old alike. They'll entertain you for hours.

Spend your vacation at Roy Acuff's Dunbar Cave Hotel and enjoy the famous southern food, prepared in the

tradition of the old south. You will enjoy the courteous, hospitable service. The cool, quiet, restful atmosphere will give you the serene carefree attitude of complete relaxation. All appointments are modern and designed for gracious living.

Dunbar Cave was what is today referred to as a Dance Cave. It would appear that dance caves flourished for two reasons. First of all, by their nature they were attractive natural amphitheaters. Even today, similar outdoor spots are still used for concerts in various parts of the country. But more importantly, dance caves provided pleasantly cool air that was not available elsewhere during the six months of warm weather here in Tennessee. Air-conditioning was not generally available until the 1950s. However, once air-conditioning arrived, concerts were more likely to be housed indoors and in town in all-weather facilities that were more easily accessible to larger numbers of people. The days of the dance caves were over. Roy Acuff's purchase of Dunbar Cave seems to be especially ill timed, since air conditioning became widely available within just a few years.

The second Idaho Springs Hotel was destroyed by fire in the 1950s and was never rebuilt. Shortly thereafter, W.W. Dunn again resigned, due to Roy Acuff's decision to sell beer at the concession stand. Spot and Gladys Acuff then became managers of the cave. Due to the growing lack of business, Roy Acuff sold the property in 1966 to a local contractor named McKay King. The eight years of the McKay King era (1966–1973) will be discussed in the next chapter.

Roy Acuff, known as the King of Country Music, died on November 23, 1992.

Figure 9-11. Roy Acuff, owner of Dunbar Cave from 1948 to 1966 and World Famous Entertainer. Photo courtesy of Billyfrank Morrison.

Figure 9-12. Roy Acuff and his son Roy Neil Acuff. Photo courtesy of Billyfrank Morrison.

Figure 9-13. Roy Acuff In Person.
Postcard from the Billyfrank Morrison collection.

Chapter 10

The McKay King Era: 1966 – 1973

In 1966, Roy Acuff sold Dunbar Cave to a local contractor and politician named McKay King.

According to the *Tennessee Encyclopedia of History and Culture*, the swimming pool at Dunbar Cave closed in 1967. One can still see the concrete outline of the edge of the pool behind the State Natural Area Office. It has been filled in with dirt and is now covered in grass.

A 1971 Guided Tour

According to caver Ernie Payne, the commercial tour of Dunbar Cave was still in operation as late as 1971. He recalls his visit:

In the spring of 1971, after taking a tour of Fort Campbell, I drove my family to Dunbar Cave. When I was a boy, I remembered passing through Clarksville and seeing the sign showing the turnoff to the cave, but my father had no interest in caves, so I didn't get to go to the cave either. Now, I was in the driver's seat and still interested in Dunbar Cave.

By the time we parked the car a light rain was starting to fall. We bought our

Figure 10-1. People in boats on Swan Lake.
Postcard from the Billyfrank Morrison Collection.

Figure 10-2. People in the swimming pool.
Postcard from the Billyfrank Morrison Collection.

tickets and waited for the next tour. This gave us a chance to look at the large wooden dance floor covered by a large roof, which was built in front of the cave entrance. There were tables and chairs arranged around the edge of the floor. We could easily imagine a dance being held here on a hot Saturday night, with the cool cave air offering relief to band and dancers alike. Then we heard the announcement over the public address that the cave tour was about to begin.

I really didn't expect much of a cave but was surprised at the passage size and the speleothems. After the usual introductory remarks, we were led into the cave and stood by the River Styx for some more history of the cave.

The best I can remember is that we entered the Counterfeiters Room to hear its story, plus a comment about there once being another way out of the cave from that room. After looking at the mannequins, who represented the counterfeiters, we wandered to the large area of stream passage and crossed a bridge.

We walked into the Indian Trails Passage as far as the drawing on the wall

of the soldier whoever he was. We then backtracked and crossed another bridge and into a passage filled with Civil Defense supplies. Of course, we heard what a "good" fallout shelter this cave was for the local people.

After going through a room with side passages, we entered Independence Hall. Here we saw colored lights shining upon a profusion of speleothems. Water drops fell upon us occasionally while we were in this room. Perhaps the light rain outside the cave provided the extra pressure to drop the water for I have been in Independence Hall more recently in dry weather and can tell the difference. This room was the highpoint of the commercial tour. From here we headed for the surface. We then sat at one of the tables along the dance floor and enjoyed some refreshments before heading back to Evansville.[1]

Sometime shortly after Payne's visit, Dunbar Cave closed forever as a private, commercial cave.

---

1   Ernie Payne, "Dunbar Cave Comments," *The Michiana Caver*, vol 5, no. 11, pp 105–106.

Figure 10-3. Many people have carved their initials on this Beech Tree over the years. Photo by Larry E. Matthews, March 22, 2011.

McKay King died in 1971, leaving the Dunbar Cave property to his widow. In 1972, the City of Clarksville purchased the 300 acres of the Dunbar Cave property where the golf course is located. This golf course is still operated by the City of Clarksville and is a very popular part of the city's park system. Then, in 1973, the state of Tennessee purchased Dunbar Cave and the remaining 110 acres of land to become a part of the State Natural Areas Program. This will be discussed in Chapter 13 – State Natural Area.

Dunbar Cave

Chapter 11

My First Visit To Dunbar Cave

Sometime in the early 1950s, when I was about six or seven years old, I made my first visit to the Dunbar Cave property. My mother and I drove from Nashville, Tennessee, to meet my aunt and cousin, who lived in Hopkinsville, Kentucky, at the time. The purpose of our visit had nothing to do with Dunbar Cave. We were to meet and go swimming in the pool. There was a very nice concrete swimming pool, complete with a bathhouse and a snack bar. The pool over-looked the 15-acre lake and paddleboats were available for rent.

That afternoon I made my way to the cave entrance. Dunbar Cave's entrance is exceptionally large and attractive, especially with the "improvements" that had been made to convert it into a dance area. The view of the lake from the dance floor is superb. There was a snack bar and there were tables and chairs. But to me, the large, dark cave entrance itself was irresistible. I walked back as far as I could to the iron gate that bars the way into the cave. Staring into the large, dark, mysterious cave passage, I thought it was the most fascinating thing I had ever seen in my life. Although it was a hot, humid summer day outside, the cool air blowing out of the cave was refreshing and inviting. More than anything else, I wanted to go into the cave and see what was there. I asked my mother if we could take a tour of the cave, but she told me that they were not giving tours that day. As it turned out it would be another 22 years before I finally got to go inside Dunbar Cave.

To the best of my knowledge, that was the first cave entrance I ever saw. It sparked my imagination as a child and my desire to go in caves never went away. And here I am, back at Dunbar Cave again.

* * * * * * * * * *

In 1961, the Tennessee Division of Geology printed Tom Barr's classic book, *Caves of Tennessee*. My friends and I were sophomores in high school that fall and we read the book from cover to cover ... then read it again ... and again. So many interesting caves and we wanted to go to them all. The description of Dunbar Cave sounded particularly interesting:

> Dunbar Cave has been open to the public for more than 75 years and is a noted recreational spot. The cave stream emerges below the mouth and is dammed up to form a lake. At present the cave is electrically lighted.
>
> The entrance is 35 feet wide and 10 feet high. The cave consists of three large, subparallel galleries and a number of smaller side passages. The large galleries trend north-northwest and average about 15 to 20 feet wide and 8 to 12 feet high. The main avenue of the cave is 2,000 feet or more in length. About 1,500 feet from the mouth is "Peterson's Leap," a 35-foot pit developed by slumping of the fill into a lower level. According to tradition, a man named Peterson became lost in the cave

and fell into the pit.

West of the main avenue is a series of large chambers along the stream, which is wide, deep, and flows slowly, almost imperceptibly, through the cave. It is inhabited by blindfishes (*Typhlichthys subterraneus*).

East of the main avenue is a series of chambers well decorated with dripstone, the most popular section of the cave. The largest of the formation rooms are "Independence Hall" and "The Egyptian Temple." Near Peterson's Leap a crawlway extends northward for 150 feet into a seldom-visited room which contains a deep, rock-rimmed pool and a large, active flowstone formation.

The main cave was explored as far as a 20-foot solution pit which extends across the floor from wall to wall. A large avenue can be seen extending beyond, but its length was not determined. The total length of passages visited by the writer is about 4,500 feet.[1]

Hmmm, a large, unexplored avenue of the other side of a 20-foot pit? We **had** to go there, cross that pit and explored that passage. But, it took us a while to actually get there.

Sometime in 1963 or 1964, several of my high school caving companions and I drove to Dunbar Cave in hopes of gaining permission to enter and explore the cave. As usual, we stopped to ask the owner's permission. To our shock, we were given the worst verbal abuse imaginable. Instead of merely saying that the cave was closed and no one was allowed to enter, or that the cave would be open for tours at some time, or even that there was a charge for admission, this person chose instead to curse and threaten us. Needless to say, we left immediately. Naturally, we had out trusty copy of Barr's *Caves of Tennessee* with us, so we had plenty of other caves to choose from. We went to another cave in Montgomery County that day. And so, my second trip to Dunbar Cave also ended without my setting foot inside the entrance. It would be 1974 before I finally made it inside the cave and across that pit.

---

1   Thomas C. Barr Jr, *Caves of Tennessee*, Bulletin 64 of the Tennessee Division of Geology, 1961, p 331.

Chapter 12

Fallout Shelter

During the height of the Cold War, the Federal Government not only considered caves for use as fallout shelters, they actually went so far as to designate a number of caves as official public fallout shelters. Some of these caves, including Dunbar Cave, were actually stocked with supplies. The theory was that if people had enough warning, they could go to one of these designated caves and be protected, first from the blast of the atomic bomb when it exploded, and then spend several more weeks protected from the radioactive fallout. In retrospect, it is doubtful that people would have had adequate warning before a nuclear attack to seek shelter in these caves. Further studies of caves showed that the considerable flow of air and water through their passages would have resulted in their being of little value, if any, for protection from fallout. As Joe Douglas wrote: "In the early 1960s, (federal) officials ignored previous studies, which outlined the problems of using caves as fallout shelters, and implemented a hastily conceived and poorly designed program, which affected hundreds of caves across the nation."[1] Joe Douglas also notes that: "It was William J. Stephenson, a government patent examiner and the founder of the Speleological Society of the District of Columbia and, in 1941, of the National Spele-

ological Society, who first brought forward the idea of utilizing natural caves as shelters from air attack."[2]

To further complicate matters, caves are relatively cold and damp. Fifty-six degrees seems delightful when you are caving, but if you sit down for more than ten minutes, you definitely start to get cold. For people to spend days or weeks in a cave, they would need to be dressed much more warmly than people in the South are accustomed to dressing. And, like most caves, Dunbar Cave is muddy. Where would people sit and sleep? What provisions would be made for restroom facilities? What would be the source of light? All of these problems appear to have been ignored in the effort to make the populace believe that they actually had a safe place to go in case of nuclear war.

According to the files in the Dunbar Cave State Natural Area Office, Roy Acuff agreed to let Dunbar Cave be used as a fallout shelter on June 29, 1965. It would be the largest fallout shelter in Montgomery County and was designated to house 2,603 people for two weeks. How the Federal Civil Defense officials arrived at such a precise figure is truly remarkable.

Not only was Dunbar Cave designated as a nuclear fallout shelter, it was actually stocked with supplies. It was one of very few caves to ever receive both the fallout shelter designation and the supplies. The supplies were stored inside an enclosure made from wood and chicken wire.

1 Joseph C. Douglas, "Shelter From the Atomic Storm: The National Speleological Society and the Use of Caves as Fallout Shelters, 1940–1965," *The Journal of Spelean History*, vol 30, no. 4, pp 91–106.

2 *Ibid.*

Figure 12-1. Remains of vandalized fallout shelter supplies in Dunbar Cave.
Photo by Tom Cussen, 1977.

Figure 12-2. Green fallout shelter water drums in Grand Canyon Caverns, Arizona. Similar drums were in
the Dunbar Cave fallout shelter stockpile. Each drum held 17.5 gallons of drinking water.
Photo © copyright by Jansen Cardy, 2010. Used by permission.

Apparently this stockpile was a source of pride and the guides would point it out on their tours of the cave. The exact contents of the stockpile are unknown, but appear to have been mostly crackers, drinking water, and first aid supplies. According to Joe Douglas, "typical provisions included tinned crackers, 17.5-gallon water cans, sanitation kits, medical supplies, and radiation monitoring devices, in lots of 50. Enough provisions were included to cover a projected two-week stay in the caves."[3]

At some point during the early 1970s Dunbar Cave was broken into repeatedly by vandals, who set the fallout shelter supplies on fire. The resulting fire left a large black soot deposit on the ceiling that is still visible today. The heat from the fire caused several large rocks to fall from the ceiling. Several of the green canisters that the supplies were stored in are still pinned beneath these rocks today. A few burned crackers were still there in 2004. The remainder of the burned debris, however, has been carried out of the cave.

Grand Canyon Caverns in Arizona still has

Figure 12-3. Civil Defense fallout shelter sanitation kit in Grand Canyon Caverns, Arizona. Similar kits were in the Dunbar Cave fallout shelter stockpile. Photo © copyright by Jansen Cardy, 2010. Used by permission.

its fallout shelter supplies. Caver Jansen Cardy has provided photos of those supplies for this book so that we can see what the Dunbar Cave stockpile must have looked like before it was burned. The Lost Sea in Monroe County, Tennessee still contains a large number of tins that contain fortified crackers.

3 *Ibid.*

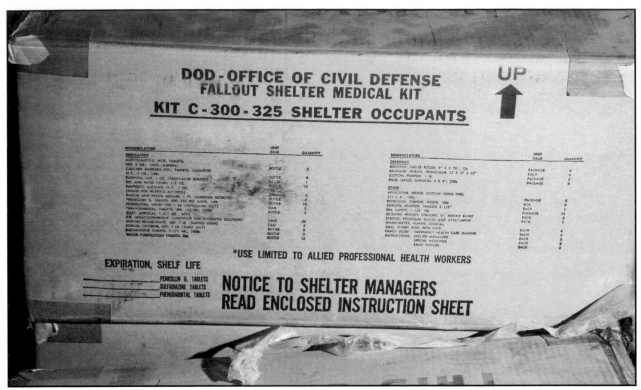

Figure 12-4. Civil Defense fallout shelter medical kit in Grand Canyon Caverns, Arizona. Similar boxes were in the Dunbar Cave fallout shelter stockpile. Photo © copyright by Jansen Cardy, 2010. Used by permission.

Figure 12-5. Tins of Fortified Crackers in the Lost Sea, Monroe County, Tennessee. Similar tins were in the Dunbar Cave fallout shelter stockpile. Photo by Larry E. Matthews, March 12, 2008.

Figure 12-6. Part of the large Civil Defense fallout shelter stockpile in Grand Canyon Caverns, Arizona. Photo © copyright by Jansen Cardy, 2010. Used by permission.

Figure 12-7. Part of the large Civil Defense fallout shelter stockpile in Grand Canyon Caverns, Atizona. Note the toilet seat and three rolls of toilet paper. Photo © copyright by Jansen Cardy, 2010. Used by permission.

Figure 12-8. Green fallout shelter water drums partially buried after the ceiling collapse caused by the fire in Dunbar Cave. Photo by Larry E. Matthews, December 7, 2004.

Chapter 13

State Natural Area

The state of Tennessee established a State Natural Areas Program in 1971. Since that time the General Assembly has designated 68 State Natural Areas. Some of these sites are owned by the state, and others are managed through Cooperative Management Agreements with other local, state, and federal agencies as well as with non-governmental organizations and private land owners.

In 1973 the state of Tennessee used a large federal grant to purchase Dunbar Cave and the surrounding property from McKay King's widow. The purchase price for the cave and the surrounding 110 acres of land was $265,500, of which half ($132,500) was Federal grant money.[1] Ironically, the caving community had been promised for several years that as soon as the state had enough money that it would buy Big Bone Cave in Van Buren County. But, despite this promise, the money went to buy Dunbar Cave instead. Big Bone Cave contained the best-preserved saltpeter mine in the United States, but without protection, it was being seriously vandalized. Fortunately, several years later, Big Bone Cave was acquired and it, too, is now a State Natural Area.

In the long run, it appears that the purchase of Dunbar Cave has been an extremely positive addition to the State Park Program. The land above the cave has become one of the most visited units in the State's Natural Areas Program. The property boasts a 15-acre lake, fed by the stream from the cave, and approximately 3 miles of hiking trails. The cave, which had been vandalized as severely as any in the state, has been cleaned up, and is a very popular destination, with the state offering guided tours on a regular basis. More people visit this cave than any other cave in the State Park System. But most importantly, over 5 miles of virgin cave has been discovered since the state bought this property, and now this New Discovery Section constitutes an important underground wilderness.

Several cavers in the Nashville Grotto of the National Speleological Society, especially your author, had wanted to explore Dunbar Cave for years, so when we heard that it had been purchased by the state and was now public property, we decided to drive up to Clarksville to see if we could obtain permission to go in. I wrote the following account of that trip in 1974:

On Sunday, June 16, 1974, Steve Loftin, Larry Adams, Cher Woodward, two of Larry's (Adams) friends from Lewisburg, Marietta (Matthews), and I drove up to Dunbar Cave in Montgomery County to cross the pit that Barr described as blocking the main passage (see Barr, p 331). Dunbar Cave was bought by the state of Tennessee just a few months ago.

1   United States Department of the Interior, Bureau of Outdoor Recreation, Form BOR 8-92, on file at Dunbar Cave State Natural Area Office. Project Number 47-00204, dated February 11, 1974.

Figure 13-1. Larry Adams and Steve Loftin in Independence Hall.
Photo by Larry E. Matthews, 1974.

When we arrived at the Dunbar Cave property the gate was open with a sign which read: "Dunbar Cave Natural Area—Caution: Under Development—State of Tennessee." Well, we knew we were going to be cautious, so we just drove on in and drove back to the old parking lot near the cave entrance. Here there was a state highway patrol car and two troopers in uniform. I just assumed that they had been assigned to police the state property. We got out of our cars, in street clothes, and started walking toward the cave entrance. One of the troopers asked us, "Where do you think you're going?" To which I replied, "We're just looking around." This seemed to satisfy him, and after a few minutes they seemed pretty friendly. The gate to the cave was open (vandals had broken the lock), but the gate was clearly posted "Keep Out—Entrance by Permission Only." Since we seemed to be acceptable to the two troopers I asked them if it would be all right if

we went into the cave, to which they said, "Sure, go ahead." This seemed to satisfy the requirement of "Entrance by Permission Only," so we went back to our cars, got our gear and entered the cave. Cher and Susan stayed with the cars.

I will not describe the cave here, since Barr's description is quite good. The cave is very large and interesting. Unfortunately it is full of trash (mostly beer cans) from the large numbers of curious visitors that have been coming from Clarksville and Fort Campbell. We saw at least 20 other people in the cave, including two little boys with one flashlight. The cave could be easily cleaned up by merely carrying out the trash, removing the old rotting wires, and removing a small amount of spray paint. Vandalism in the form of writing on walls with carbide lamps is very rare, and we did not see anyone else in the cave with a carbide lamp. There was also a large pile of empty drums, which were part of a fallout shelter. The cave will probably be an interesting addition to the state's park system if it is carefully developed.

We located the pit without too much trouble. A muddy but adequate ledge went around the left side of the pit and we did not have to use any of the gear that we had brought in with us. The main passage continued on the far side of the pit with its dimensions of about 20 feet wide and 20 feet high for about 150 feet. At this point it ended in a massive breakdown. We tried several paths into the breakdown, but could not break through it. We were a little disappointed, to say the least. We turned around and did a little sightseeing through the rest of the cave and took some pictures in Independence Hall, the most attractive formation room in the cave.

As we left the cave and walked up to the cars, I noticed that there were now *two* police cars. Cher informed us that we were

all "under arrest."

I asked, "What for?"

"Trespassing."

As this point we learned that these were Clarksville City Police, so I explained to them that the two state troopers had given up permission to enter the cave. The fact that the troopers were no longer there worried me, and when the Clarksville policemen wanted to know the names of the two troopers (which, naturally, I never had thought about getting) I really was feeling badly. The police were surprised when I told them that there were quite a few other people still in the cave. Maybe we had honest faces, or something, because after a few minutes they told us they would let us go if we would leave immediately. You have rarely, if ever, seen people throw caving gear into cars and leave so fast.

The moral of this story, I suppose, is that we should have gotten *written* per-

Figure 13-2. Dunbar Cave State Natural Area Office.
Photo by Larry E. Matthews, December 7, 2004.

mission, instead of verbal permission. We assumed that the state troopers would still be there when we came out, which was a bad assumption, as it turned out. At the very least, we should have written down the names and license number of the two troopers.[2]

In the above article, I made the comment that: "The cave could be easily cleaned up by merely carrying out the trash, removing the old rotting wires, and removing a small amount of spray paint." I also said that: "The cave will probably be an interesting addition to the state's park system if it is carefully developed." For whatever reason, little, if anything was done to clean up Dunbar Cave for years. An article by Nashville Grotto member David Stidham describes a clean-up trip (by cavers, of

course) that took place on February 24, 1979, nearly five years later:

This was a joint trip of the Nashville Grotto and the Nashville Chapter of the Tennessee Trails Association, with 15 adult and youth members of Girl Scout Troop 127 as special guests of the Tennessee Trails Association. About a third of the participants on the trip had never been in a cave before, about a third were still novices, and the remaining third were experienced cavers. With Larry Matthews in the lead and David Stidham bring up the rear, the experienced cavers placed themselves among the novices throughout the trip through the cave.

David Stidham had made prior arrangements with the park ranger for the Dunbar Cave State Natural Area to take the group into the cave. The unusually large group was met at the entrance by Paul

2   Larry E. Mathews, "Fun And Games At Dunbar Cave," *The Speleonews*, vol 18, no. 4, pp 54–55.

Figure 13-3. Sign at the entrance to the Dunbar Cave State Natural Area.
Photo by Larry E. Matthews, December 7, 2004.

Figure 13-4. The office for the Dunbar Cave State Natural Area was once the bathhouse for the swimming pool. Compare this photo to Figure 9-4 and 10-2. Photo by Larry E. Matthews, March 22, 2011.

Huff, the state caretaker, who unlocked the gate. Once in the cave, a bar was placed on the door to prohibit other unauthorized groups from entering. When the entire group exited the cave that afternoon, the trip leader placed the lock back on the gate.

Before entering the cave, the trip leaders briefly discussed proper safety and cave conservation practices. The history and significance of the cave was discussed by Larry Matthews, and Mr Huff said a little bit about the archeological investigation currently in progress, as evidenced by the dig visible in the mouth of the cave. Litterbags were then distributed for litter picked up during the cave trip. Larry led the group through the Main Passage to the Junction Room, stopping briefly at the Counterfeiters Room. From the Junction Room, the group went through a dug crawl to the Hidden Lake Passage. Backtracking, the group then traveled through the Great Relief Hall to the beautifully decorated Independence Hall. Time was spent there enjoying the formations, photographing, exploring the Crystal Palace, and eating lunch. After lunch, a little time was spent in the passage off Independence Hall. Then the group was led to the Stone Mountain Room.

Returning to the entrance via the Main Passage, most of the Tennessee Trails Association members and guests exited the cave. Litter, which was picked up in the cave, was deposited in garbage cans at the parking lot. The remaining cavers returned to the cave to haul out larger pieces of trash. Removed from the cave were several hundred feet of old metal pipes, several large wooden signs, miscellaneous pieces of old electrical equipment and posts, and

Figure 13-5. The Dunbar Cave Entrance looks lonely and deserted compared to its days as a resort. Compare this photo to Figure 8-8. Photo by Larry E. Matthews, February 15, 2005.

many garbage cans and metal containers which were part of the Civil Defense project. Literally tons of the Civil Defense junk was left in the cave for a more organized clean-up trip. It is recommended that the Nashville Grotto organize a full-scale clean-up trip in the cave, using wheelbarrows, and thoroughly clean up this mess. Graffiti should also be removed from the walls.[3]

For whatever reason, after five years of ownership, the state of Tennessee had done little, if anything, to clean up Dunbar Cave. In fact, the cave was in much worse shape than when I had visited there in 1974. It would take more involvement by organized cavers in the years

to come to clean the cave up to an acceptable level. The dirty little secret was that the state of Tennessee had squandered federal money, giving it to politically-connected people to buy property that they had absolutely no interest in. It took local Clarksville residents with a love for the cave and the caving community to force the state into actually making the property clean and usable. Your tax dollars at work!

On July 28, 1979, Nashville Grotto cavers Joel Buckner, Mark Miller, John Hoffelt, and David Parr conducted another clean up trip to Dunbar Cave. Joel Buckner describes the trip as follows:

We came equipped with assorted tools (wheelbarrows, scrub brushes, shovels, saws, and the like) and were able to accomplish quite a bit. We used the wheelbarrows to bring out the large (and sometimes

---

3   David Stidham, "Dunbar Cave State Natural Area," *The Speleonews*, vol 23, no. 2, pp 34–35.

heavy) Civil Defense cans. At least a ton or more of those were removed along with several very large sheets of plastic. All of the remaining water pipe in the cave was removed with the aid of hacksaws. Sections of pipe that were a couple of hundred feet long or so were cut into smaller pieces which made the load lighter and also made it possible to negotiate the curves in the passage on the way out. All of that took several hours after which we scrubbed graffiti off the cave walls and picked up scattered trash.

Although personnel on this trip was limited, substantial progress was made in cleaning up the cave. One more well equipped trip, hopefully with a few more people, would probably take cave of everything still in the cave.[4]

* * * * * * * * * *

Unknown to the Nashville Grotto cavers, a group of organized cavers from Indiana was also visiting Dunbar Cave at this time. Their

Figure 13-6. A sign shows the trail system at Dunbar Cave State Natural Area. Photo by Larry E. Matthews, March 22, 2011.

story, told by Richard P. Geer, sounds familiar:

Pathetically, in the two years between McKay King's death and the state purchase, local vandals made a shambles of the elegant bathhouse and concessions building and broke down the cave gate. Civil Defense supplies were destroyed and strewn about the cave in a trail of wreckage that missed few passages. The fine lighting system was torn down and strewn about the cave, and the walls were attacked with

4   Joel Buckner, "Dunbar Cave Clean-Up," *The Speleonews*, vol 23, no. 6, p 118.

spray paint. It is a miracle that the formations were not ruined, yet these too were vandalized.

Joe McCraw and Paul Huff were hired as the fledgling park's first staff. They spent the next three years attempting to drive local vandals from the cave and initiate a park and cave clean-up program. They repeatedly repaired the gate—almost daily for a time.

By the time the first of Northern Indiana Grotto's cavers arrived on the scene that March night in 1977, the gate had been secured to a degree and, as Joe stated, "We cleaned up a lot ... hauled out three truckloads last fall."

Joe accompanied us on the first trip and on many others that year. As the heavy iron door was swung open, the cool, damp, musty air blasted strongly against us. Underneath the normal cave smells we could sense a dank, moldy stench, like rotting ... what? It was difficult to identify the metallic odor. The passages were filled with trash—broken chairs, smashed tables, and bent and twisted electrical conduit. There were shattered light fixtures and wad after wad of kinked and twisted wire. Rusty, olive-drab, 25-gallon drums were scattered about, contents strewn about in disorder. In the entrance room were two heaping piles of unidentifiable garbage.

It was difficult to believe what we were seeing, yet as we progressed deeper into the cave the accumulation of trash and garbage became even worse. Once, after an especially long silence, Joe McCraw asked a simple question of bewilderment, "Why do you suppose anybody would want to do something like that?" He didn't really expect an answer, and we had none.

As we began the ascent to the Junction Room we passed an elliptical-roofed tunnel which was filled halfway to its charred and blackened ceiling with garbage, trash, and burnt and broken Civil Defense canisters once representing the life support for perhaps hundreds of people. The stench was sickening.

"What was that?" I exclaimed.

"Oh, that?" That's where the fire was!" Joe answered, "We haven't gotten to that yet."

"Fire? What fire?"

"Oh, they set a fire up here, burned for a week, whole cave was filled with smoke, nobody could get in to put it out. Guess it smothered itself."[5]

The cavers from Indiana must have wondered why on Earth the state of Tennessee would buy a cave and leave it in such deplorable condition. Ironically, it was clear that the cave was in much worse shape by 1977 that when the Nashville Grotto visited there in 1974.

* * * * * * * * *

When you read the above accounts of the massive damage that happened to the Dunbar Cave property *after* it was purchased by the state of Tennessee, you can only conclude that the people that were entrusted by the taxpayers to protect and oversee this public property were criminally negligent in their duties. This was politics at its worst, the squandering of valuable federal grant money to buy property that the state did not even want, just to pay off their political buddies. Then, sitting back and letting the property be destroyed without raising a finger. How simple would it have been to protect the cave? If they could not control access, they could have simply welded the cave gate shut until such time as they had the money and personnel to protect the cave. But they didn't. Why? Because they didn't care. In addition to countless illegal acts of trespass and property destruction, which were in violation of local and state laws, the destruction of the fallout shelter supplies probably also violated federal laws.

---

5   Richard P. Geer, "The Showplace of the South: Dunbar Cave," *NSS News*, vol 38, no. 10, pp 224–229, 232–233.

Chapter 14

The Indiana Connection

Dan McDowell was stationed at Fort Campbell, Kentucky, from February, 1962, through August, 1964, as a member of the 101st Airborne Division Support Group. The Support Group Sergeant Major was Paul B. Huff, and as a result of some work assignments for him, Dan came to know Huff well enough that he kept in periodic touch with him after he was discharged from the Army.

While stationed at Fort Campbell, Dan visited his first wild cave, Cooper Creek Cave in Montgomery County, Tennessee. This visit occurred in August of 1963. Also while he was stationed at Fort Campbell, Dan had made several visits to Dunbar Cave, which had the best public swimming pool in the Clarksville area. Dan married a local girl from Clarksville, Tennessee, and that established local family connections. For several years, until the early 1970s, all Dan's cave exploration was in the Clarksville area with family and friends. Dan recalls his early caving career:

> I joined the NSS and organized caving in 1970 with the Windy City Grotto in Chicago, Illinois, and then in 1974 I joined the Northern Indiana Grotto in Mishawaka, Indiana. During this time I still caved much more in Tennessee than in Indiana. I developed an interest in finding new caves (that is caves not in Barr's or Matthews' books on Tennessee caves) in Montgomery and nearby counties. I spent a lot of time pursuing leads

and talking with numerous people. I did find quite a few new (to me) caves. Many of the local people made reference to Dunbar Cave, which I had visited in 1967. An old Chain Gang Warden told me of several work details he led for a couple of weeks excavating some of the trench-ways in parts of the cave back in the early 1930s.

> I knew Dunbar Cave had been closed in the early 1970s. I was driving by the cave one day in the fall of 1976 when I saw a state truck in the parking lot. I stopped and met Joe McGraw, who was the first caretaker the state hired to manage the property. Joe was really pleasant and informative, but told me the cave was not open to visitors yet, let alone cavers. Near the end of the visit, almost in passing, he mentioned that Paul Huff, a retired Sergeant Major, was the other caretaker. I nearly swooned. It had been three years since I last saw Paul, but I knew immediately, I'm in the cave. Joe told me that if I knew Paul, I'd probably get to go into the cave. I told Joe I was off to see the Sergeant Major, and he laughed. A couple days later, on a Saturday morning, my brother-in-law and another friend and I took a three-hour trip in the cave. Man, is it neat to have the run of a former commercial cave.

> After a second trip, a couple of weeks later, and after talking with Paul and Joe,

Figure 14-1. *The iron gate at Dunbar Cave Entrance. Photo by Tom Cussen, 1980.*

I found out the only map of the cave they had was an old Civil Defense line survey done by Frank Meador showing about 2,000 feet of passages. I offered to see if my Grotto (Northern Indiana Grotto) would be willing to map the cave if the state would grant permission. Joe and Paul talked with officials from the Tennessee Department of Conservation and shortly, around the end of the year, I received a letter of permission from Paul Huff to map the cave. The Northern Indiana Grotto discussed and approved the project at the January and February, 1977, meetings. The first survey trip was on March 12, 1977.[1]

* * * * * * * * *

Richard P. Geer, a caver with the Northern Indiana Grotto, recalls his first trip to this Tennessee cave:

It was a dismal, rainy night. As often seems to be the case with the carefully laid plans of cavers, ours had inadvertently gone awry, there was no one waiting there at the chained and padlocked double gates to allow us entry. We knew we were at the right place though, for the rustic sign hanging above the wide portal proudly proclaimed "Dunbar Cave State Natural Area — Under Development." Braving the falling drizzle, we abandoned our vehicle and explored a ramshackle white building that stood out in stark contrast to the black night. To our surprise, we found it had once been an elegant bathhouse, and a gaping cavity behind it had been a fine concrete and mosaic swimming pool.

Dan McDowell had visited the cave in 1976 at the invitation of two old Army buddies from Clarksville, Joe McGraw and Paul Huff. They had been assigned

1 Dan McDowell, Personal communication, December 7, 2004.

Figure 14-2. Training new surveyors. Shown left to right: Al Goodrich, Steve Couts, Dave Collins, and Walter Scheffrahn. Photo by Tom Cussen, 1977.

to the first staff of the fledgling park. For all the cave's long history, Joe told Dan, there was no map. Dan offered to have Northern Indiana Grotto map the cave. A splendid idea, despite the 900-mile round-trip drives. Had we realized what the project would become as we sat there in the rain that dreary night in March, trying to grab a bit of sleep against the staccato pounding of the rain against the roof, I rather doubt that we would have taken on a project so far away from home. But the next day I saw the entrance.

Nestled into a high limestone bluff at the head of the mud-stained lake was a glittering arcade standing in the morning sun like the ruin of an ancient temple, whispering a breath of mystery with each breeze. I stood spellbound by the fascinating sight. This was "The Most Beautiful Cave Entrance in America."

On that first visit in March, with Walter Scheffrahn in charge of cartography, we began the survey from the cave's old iron gate. Al Kurta, Jack Countryman, Dan McDowell, and I formed a group of the grotto's more skilled surveyors to begin mapping major passage to the east while Walter kept less experienced mappers in the main passage.

Altogether 3,850 feet were mapped during this first trip. We were elated by our success. We were committed. For better or worse we'd make the frequent trips to Clarksville.[2]

Dan McDowell describes their relationship with the state:

It was amazing how supportive the state people were. Besides Joe and Paul giving us the keys to the cave, we could drive up to the entrance, take our camping gear in, and stay the whole weekend in the cave. Some of us wore house slip-

2 Richard P. Geer, "The Showplace of the South: Dunbar Cave," *NSS News*, vol 38, no. 10, pp 224–229, 232–233.

117

Figure 14-3. Dave Collins, Jack Countryman, Alan Goodrich, and Walter Scheffrahn in Independence Hall. Photo by Tom Cussen, 1977.

pers while in camp. We went to Nashville to meet several Department of Conservation officials. They provided us with anything we wanted, among which was the production of 200 copies of the finished map. We thoroughly enjoyed ourselves during the project.[3]

In the next chapter we will look at the interesting details of this mapping project.

---

3   Dan McDowell, Personal communication, December 7, 2004.

Chapter 15

Modern Exploration

In the March, 1977, issue of *The Michiana Caver*, Dan McDowell and Richard P. Geer proudly announced that the Northern Indiana Grotto had officially decided to map Dunbar Cave. They mentioned that the cave was supposed to have been mapped years ago, but that the only two maps known to exist had been lost in the estate of the former owner. Before the mapping even began, Dan McDowell and Richard P. Geer estimated how long the project would take:

> We presently have no idea of the cave's actual size, but from old brochures, and research from other areas, we visualize some 8,000 to 9,000 aggregate feet of passage in a multi-level system. This compares favorably with current knowledge of caves in the Clarksville area. There is a sizable stream in the cave, the River Styx, which is apparently not seriously sensitive to flooding. The river contains blind cavefish. The upper levels contain several formation rooms. Independence Hall features several large flowstone columns. The Egyptian Grotto is another of the cave's popular features. Petersons Leap is a 35-foot pit within the cave, also popular back in the cave's commercial days. Much of the cave is accessible via old tourist trails, and up to 200-foot shots can be taken in some areas. Off the tourist trails are some outstanding formation rooms, which are not vandalized. We imagine that with such

a long history as Dunbar Cave has, the cave has been quite thoroughly explored. The prime objective will be to accurately map the cave. Of course, one never really knows what might have been overlooked until the job is begun.

> Due to travel time and distance, the three or four trips necessary will be strictly work trips. Those attending will be prepared to spend a minimum of 18 hours on the road, and a long stay in the cave. We may even camp in the cave. Since the cave is securely gated, excess gear can be stored inside without fear of theft; this is a treat in this day and age. It will be necessary to take in extra provisions and clothing. The temperatures of Tennessee caves are slightly higher than are found in Indiana. Thus, Dunbar Cave will be a bit more hospitable than you'd expect in Indiana.[1]

The first trip to begin mapping the cave was scheduled for the weekend of March 11, 1977. Walter Scheffrahn describes that first mapping trip:

> We "slept" in our cars until Dan McDowell arrived early Saturday morning with the park attendant, Joe McGraw. Joe unlocked the gate and led us into the park

---

1 Dan McDowell and R.P. Geer, "Dunbar Cave—a new project," *The Michiana Caver*, vol 4, no. 3, pp 19, 26.

Dunbar Cave

Figure 15-1. Surveyors camped in the Junction Room. Shown left to right: Allen Kurta, Walter Scheffrahan, Steve Couts, Dave Collins, and Pat Harvey.
Photo by Tom Cussen, 1977.

grounds. We were prepared for a stay in the cave, and thus headed into Dunbar.

We selected a base camp about 700 feet inside the entrance. Here we unpacked and ate breakfast. Then Dan gave us a quick tour of the cave to give us some idea of what lay ahead of us. Much of the cave is walking passage, and trenches have been dug through the crawlways. The cave was commercialized at one time and these dug passages are part of the cave's commercial development. Seeing the large rooms and the huge formations excited us all.

Back at camp, we sized up the situation and after a short strategy session we referenced our beginning survey station at the cave's iron gate. This would, so Joe explained, be the one point in the entrance that would not change.

As we started surveying inward from the gate, we began training the several non-initiated members in the group. Nine had

shown up for the trip. Because it is an 8- to 9-hour drive from northern Indiana to here, and because our objective was to map the cave, we had to teach those who had never mapped before enough so that they wouldn't be standing around doing nothing. After the first few stations, we split up into two groups. Dick, Al, Jack, and Dan went in the first team to map the remainder of the main trunk. I took the second team consisting of Erika, Liz, Mark, and Gary. We surveyed the side passages off the main trunk. I continued with the survey training.

That evening after supper, the groups from my team sacked in for the night. I replaced Dan in the first team, which was working in Independence Hall. We established "permanent stations" at the room's eight-odd side passages, and finally mapped one of the main side passages off Independence Hall. It must have been around 3:00 A.M. when we finally quit for

Figure 15-2. Steve Fisher climbing a cable ladder out of the Bat Bone Passage. Photo by Tom Cussen, 1978.

of March. Incidentally, we surveyed a total of 3,850 feet on this trip.[2]

At this point, the project was right on schedule. Two more survey trips with this much cave surveyed, and they should have the whole cave mapped. Walter Scheffrahn goes on to describe the next survey trip on March 26:

More people showed up at Dunbar Cave than I expected this time. It now seems certain that we will have no trouble finding volunteers to go to Dunbar. Members present were Dan McDowell, Al Kurta, Pat Harvey, Tom Cussen, Alan Goodrich, Jack Countryman with two friends from Oakland City, and me. We had enough trained people for two survey teams. I don't think we

the night.

The next morning we broke camp, heading out into a warm and sunny Sunday morning. The only surveying done on Sunday was from the gate out into the expansive entrance. We discussed our experiences and planned a trip for the 26th

surveyed as much this time as last, but we concentrated on Independence Hall. The passages there are beautiful, but were difficult to survey. Tom Cussen has

2 Walter Scheffrahn, "The Dunbar Trips," *The Michiana Caver*, vol 4, no. 4, pp 31–32.

*Figure 15-3. Dan McDowell and Don Hehring surveying the River Styx.
Photo by Tom Cussen, 1977.*

taken some 70 slides which he plans to show at the May meeting.[3]

After this trip, they estimate that they have mapped over a mile of passage. The next survey trip occurred on April 16 and was reported by Pat Harvey:

An uneventful 8-hour drive brought us to the gate of Dunbar Cave Natural Area at 12:30 Saturday morning. Since the caretaker was not expected to appear with the key until 8:00 A.M., we spent the night under the concrete deck at the cave entrance.

Together at last, we entered the cave after a short struggle getting Alan Goodrich and his backpack through the cave's gate, and proceeded to the base camp near Petersons Leap. Then we gave Doug and Ken a quickie tour of the cave, which developed into a re-survey of some parts of B-survey.

After lunch, Ken, Al, and I went to sur-

vey the lake below Independence Hall and the Rocky Mountain Route, the passage leading to the back side of Lovers Leap, the real Petersons Leap. A branch of this passage also leads to a series of pits. We surveyed to where the main passage makes a left turn, and then came back to investigate the large pits that almost block the last part of the passage.

Next morning, the intrepid, exhausted explorers informed us sleepyheads that we have mapped only an estimated one half of the cave. Since we've already surveyed approximately 8,000 feet that means there is a good mile to go yet. So, anyone who has not been able to come down yet should not worry about running out of cave before their chance comes. It's really worth the trip. As Ken said while we were returning from his first trip to the cave: "It's a part of me, now."[4]

---

3 *Ibid.*

4 Pat Harvey, "In The Pits—The Trip of April 16," *The Michiana Caver*, vol 4, no. 6, pp 47–49.

The next surveying trip took place on May 14, 1977. Dan McDowell describes that trip:

About 8:30 in the morning we got together to plan our work for the day. We made up two survey teams. Al, Alan, Pat, Ken, and Tom made up a team to map the breakdown room and the route through it and back to Petersons Leap. Jack, Steve, Dave, and I formed the second team, and we were to survey the River Styx from the breakdown room to the Entrance Room, and as much of the Domes Passage as we could.

We surveyed the river first. Earlier in the morning, Don and I had waded and crawled through it to see what it was like and if there were any leads off the river. This section of the river is about 600 feet long, with water depth averaging about 2 feet and ceiling height from 2 to 6 feet. Since I had already been through, I was able to stay out of the water, because of a passage that runs adjacent to the river for a couple of hundred feet from the upstream end. Downstream, a person can go for a couple of hundred feet on a low mud bank on one side. I was taking notes. They called information out to me then I walked around to the other end, about 800 feet to the Entrance Room, and continued till they emerged. Both the notes and I stayed dry. We tied in the survey at an iron rail in the Entrance Room.

In the afternoon we started surveying into the Domes Passage. This is a major passage in the cave. We surveyed about 780 feet into the main passage, and surveyed the small passages to the domes. One of those domes is about 35 feet high the other is about 50 feet high. Three other small leads were surveyed. One was a lower level back to the River Styx. We surveyed a total of about 1,200 feet in the Domes Passage, and there are maybe five more leads to survey. A couple of these will make 500 feet between them.

Just making a guess, I think we have about 2 miles of actual passage mapped. Unless one of the Domes Passage leads open up into more cave, we can finish the survey in maybe one or two more trips, and we will wind up with around 2 ½ miles of passage. We did not find any passage off the River Styx except for one small connection back to the Domes Passage. I hope that the leads in the Domes Passage will lead us to more cave.[5]

After four survey trips, Dan is worried that they are running out of cave and that the project will soon be over.

The next survey trip reported in their newsletter occurred on June 18, when Al Kurta, Ken Laughlin, Cindy Barrett, and Mark Ellis surveyed 500 feet of passages off the bottom of the Petersons Leap gorge.

Another survey trip occurred on June 25, 1977. Dan McDowell describes this trip:

On the weekend of June 25, Al and his crew were back. Jack Countryman, Dave Collins, and Steve Couts joined them, and of course I was there. The highlight of this trip was the discovery of 928 feet of virgin passage off a crawl at the top of an 18-foot flowstone wall at station B-8. I believe the passage will be called the Bat Bone Passage because of a large number of bat skeletons present there. Entrance to the passage is via a crawlway at the top of the flowstone wall, about 15 feet to a small hole that opens up for 6 feet to another small hole. It drops 4 feet to a small balcony ledge on the side of a 29-foot pit, which must be rigged. A cable ladder was tied at the top edge of the pit, but it was very awkward to get onto the ladder at this point. We climbed down a crevice which was adjacent to the pit for about 8 feet, and got onto the ladder on a free hang off the wall. About 4 feet above the bottom of the pit is a sizeable passage.

---

5 Dan McDowell, "Trip of May 14," *The Michiana Caver*, vol 4, no. 6, pp 49–50.

Figure 15-4. Sam Allen (NSS # 155) and Dan Nehring in the Indian Trails Passage.
Photo by Tom Cussen, 1977.

It is from 3 to 15 feet high and goes for about 350 feet or so to another pit that is 20 feet deep at the bottom of two climbable 10-foot drops. The total depth is about 40 feet. Dave Collins dropped the final pit and reported some standing water but no leads.

A side lead off the main passage leads to several hundred feet of maze-like cave containing domes. This section has no formations to speak of, and is very muddy. It has lots of breakdown and some of the finest clay I have ever seen in a cave. There are large numbers of exposed fossils in the walls. Someone remarked that this section looks like an entirely different cave. I agree. It is vastly different in appearance from the rest of Dunbar Cave. We are quite certain that we are the first to enter this passage. Al and Dave said there were no footprints anywhere they went except for their own. It is hard to imagine virgin passage in a cave like Dunbar, considering its long

commercial history and the long period of exploration by others. But, the extension is located in an obscure place, and there were a couple of very tight holes in addition to the pits. All of the group made it through except Jack and me. After two hours of work, Jack was able to squeeze through, but since I am the "grotto porker," another hour's work was required before the holes were enlarged enough for me to squeeze through.

Going over the survey notes and eliminating some duplication of footage necessary for accurate surveys, we have about 14,700 feet mapped—about 2.8 miles. Walter has complete figures, but this will be pretty close.

How much cave remains? There are several small leads off the Domes Passage yet, and a couple of others in other parts of the cave. We may be near the end, but again we may yet find more cave. After all, a very unpromising hole in the B survey led

Figure 15-5. Caver descends a pit using a rope. Photo by Tom Cussen, 1977.

into over 900 feet of virgin cave. Another thing to keep in mind is that so far most of the passage mapped is in the spur of the hill between the hollow in which the entrance is located and the adjacent hollow to the east. We are not really back under the sinkhole plain to the north. If we could get back under that, there would be all kinds of possibilities for a rather extensive system. Beyond the initial mapping, this is something to work for.[6]

The survey of what we now refer to as the Historic Section was completed on January 12, 1978, with a total surveyed length of 16,145 feet of cave passages (See the map at end of Chapter 5). Interestingly enough, although the cave had turned out to be nearly twice as long as Dan McDowell and Richard P. Geer had estimated, they had finished the entire project in just under a year.

6   Dan McDowell, "Progress at Dunbar Cave," *The Michiana Caver*, vol 4, no. 8, pp 63, 65–66.

*Figure 15-6. Sam Allen (left) and Don Nehring (right) in Spray Hall Dome.
Photo by Tom Cussen, 1977.*

Chapter 16

The River Styx

By 1961, when Tom Barr's book, *Caves of Tennessee*, was published, all of the more easily accessible passages of Dunbar Cave had been explored. Although more than 3.3 miles of passage had been explored in the Historic Section of the cave, it was very obvious to cavers that the cave had to be much, much longer. The cave stream, the River Styx, was a major underground river, but it could be followed upstream for only a thousand feet from the entrance to where it disappeared at the River Styx Siphon. This is the point where the cave roof lowers to the level of the water. By comparison with similar-sized caves streams, you would expect the River Styx to continue upstream for several miles and for there to be tributary streams entering along the way. Furthermore, the well-developed sinkhole plain above the cave further supported the theory that the cave stream was supplied by surface drainage into these sinkholes over an area of at least several square miles. The cave system that collected all this water to form the River Styx should be quite large indeed.

On April 30, 1978, several Indiana cavers were led on a tour of Dunbar Cave by fellow caver Jack Countryman. Jack noted that the lake was 2 feet lower than usual, since several boards had been removed from the outlet gate in the dam. Therefore, he made a point of following the river upstream to see if a lower water level in the lake would allow them to pass through what was normally the upstream terminal siphon. To his disappointment, the passage was still completely water-filled. He did observe, however, that there was clear, open water below the ledge spanning the passage, leading him to believe that it would be possible to continue upstream using scuba gear.

Upon returning home, Jack contacted Clarence "Bud" Dillon of the Bloomington Indiana Grotto about the possibility of getting him and his diving partner, Steve Maegerlein, to attempt a dive into the siphon. They talked about the proposal further at the Bloomington Indiana Grotto meeting on May 11, 1978, and Bud and Steve decided to attempt the dive on May 14, 1978.

At about 10:00 A.M. on May 14, 1978, they carried the diving gear into the cave. The two divers entered the river at the first rock abutment across the passage at the boundary of the Stone Mountain Room. From there, they swam up the river to the siphon and attached their safety line to a rock on the breakdown room side of the river, about 10 feet from the siphon. The two divers disappeared into the siphon at approximately 11:00 A.M.

Jack sat there, watching their lights disappear, wondering what they would find. When they were gone for over an hour, he became alarmed. Then two hours passed. Were they trapped? Had they drowned? They couldn't possibly have that much air in their tanks, could they? Jack recalled his dilemma:

> My real panic, as the divers were gone that trip, was that I had no idea whom to

call to get them help, especially help that was close enough at hand to be of any use. I didn't know any cavers local to Dunbar Cave and certainly not any cave divers there. Calling folks in Indiana or beyond would mean they would not be available and on-site in time to do any good if the divers were trapped, out of air, or something worse. If that were the case, I had no idea whom to call for rescue, either.[1]

And time dragged on. Three hours passed. Finally, after 3½ hours, Jack saw their lights appear in the siphon. The two divers emerged and excitedly began to describe their discoveries. Jack recalls their description:

The siphon is 350 feet long, with the passage's bottom 7 feet below the water surface, leaving 5 feet of open, water-filled passage. The passage is 7 to 10 feet wide and the floor slowly rises until it is about 2 feet below the ceiling at the end of the siphon. The ceiling slowly rises for 30 feet until there is 2 inches of air space. From

Figure 16-1. "Nestled into a high limestone bluff at the head of the mud-stained lake was a glittering arcade standing in the morning sun like the ruins of an ancient temple, whispering a breath of mystery with each breeze." Richard P. Geer, 1977. Photo by Larry E. Matthews, March 22, 2011.

there, the ceiling rises more rapidly to a point about 500 feet in where the passage is 30 feet high. (Here, Bud and Steve anchored their safety line and left it in place as a guide for future groups.)

Shortly past the end of the siphon, there is a dome on the left, and 50 to 100 feet beyond it, at the end of the 500-foot safety line, the passage makes a sharp turn to the left. Up to this point, the passage

---

1 Jack Countryman, Personal communication, November 3, 2004.

trends north-northeast. From here, it goes north-northwest. There is a small room on the right of this turn, which has a mud bank with fresh raccoon tracks on it. These tracks were noticed at other points along the passage as well. The room also contains a waterfall dropping from an upper level to the stream level.

The passage continues north-northwest as a down-cut stream channel canyon 15 to 30 feet high, 8 to 10 feet wide, with breakdown. It is wide at the top, narrow on the lower level. The lower level of the passage is muddy and carries the stream while the top part is dry and has formations along the left wall.

After 100 to 150 feet, the passage becomes side-cut, 50 to 60 feet high, and filled with large breakdown. The break-

down causes the stream to pond behind it, necessitating crawling over and through it to continue.

After another 200 to 300 feet, the passage becomes slanted vertically with the top of the canyon leaning to the left. There is a shale layer visible 35 to 50 feet up in the passage with water coming off this layer at many points. Some formations are seen along the left wall. The passage gets higher as it continues and domes are found along it, many with chert deposits. Bud reports that he traveled about 30 to 45 minutes down passage like this until he finally turned back in continuing passage and started back to rejoin Steve and come out. He reports "an uncounted number" of streams entering, many high in the passage, some lower, and many domes, most

Figure 16-2. Looking out the Entrance to Dunbar Cave.
Photo by Larry E. Matthews, March 22, 2011.

of which have enterable passage leading off. He estimated that he'd seen 4,500 to 5,500 feet of main stream passage, not counting side leads, domes, and other features, with a potential for much more to be found, and probably other entrances.

Bud and Steve noted that the passage beyond the siphon apparently floods to the ceiling for some distance. No signs of previous visitation were seen.[2]

With this trip, a new era of exploration had begun in Dunbar Cave. In one trip, they had discovered and explored a mile of virgin passageways, with many leads left to be checked. Jack's expectations of large cave past the siphon had been well founded.

2   Jack Countryman, "Dunbar Cave—the last trips," *The Michiana Caver*, vol 5, no. 6, pp 69–71.

Chapter 17

Roy Woodard

Feeling certain that it would be possible to find a new entrance that would enable them to bypass the River Styx siphon, Walter Scheffrahn, Jack Countryman, and Al Goodrich went back to Dunbar Cave on Memorial Day, 1978. Walter had received a telephone call from Dan McDowell giving him the names of some people to contact in their search for a back entrance. On May 27, they learned the name of the owner of a large, trash-filled sinkhole in the Dunbar Cave area. They were assured that this sinkhole would be the best candidate for a back entrance. Walter describes their trip as follows:

We contacted the owner, and after explaining our intentions and purpose to him, he agreed to allow us access, and even permitted us to camp on his land. We set up camp right next to the sinkhole's perimeter and prepared to enter. We had no idea what to expect.

We descended through 40 feet of junk cars and other debris, finally coming to a small alcove. In the left wall we found a small crack, which we squeezed through into a crawlway. The crawlway zigzagged three times, dropping some 5 feet, then ended. To one side was a gravel-covered squeeze way. Jack had to clear away some of the gravel so he could get through. It gradually opened up until we entered a series of four adjacent domes, some 20 feet high and 10 by 30 feet in size. There was a 5-foot drop to the floor of the first dome. Imme-diately to the left was a stoopway some 40 feet long, 3 feet high, and 12 feet wide. It descended at a grade of about 1 foot in 5.

The stoopway opened into a large, gaping hole. Al went back to camp for vertical gear and rigged and dropped the pit. He described it as being some 30 feet deep with a base of 20 by 30 feet. Two large passages diverged from the pit. Al explored some distance into the larger of the two passages and reported back that there was plenty of cave.

We then went back to camp where we had a good meal and made plans for Sunday. Then we got a good night's sleep.

After breakfast we gathered our vertical gear together and went back into the cave to drop the pit. We checked out both passages. One ended in two large domes. The other passage was 6 feet by 8 feet and reminded me of a dry riverbed. It meandered for several hundred feet and finally entered into a stream passage. I had brought along a compass and took compass readings along the way, so we could get some feeling of the passage's direction. At the stream passage junction we turned downstream in the hope of finding some evidence of where Bud and Steve had been on their dive.

The stream passage varied greatly in size, from 8 feet to over 30 feet high. Several large rooms were encountered and the canyon passage was huge. We found at least half a dozen side leads, which we didn't bother to enter, saving them for a later date. The stream varied in depth from ankle deep to over waist deep. The mud was fine and silty and every step we took, we sank in over our ankles, which made walking difficult. Finally, we crawled through a break-down area where we found much loose rock. On the other side of this obstacle we were wildly overjoyed to find the footprints of one of the divers. We now had no doubt that we were in the stream that becomes the River Styx. We pushed on until we were in shoulder-deep water and the ceiling was descending. Assuming that we were near the siphon head, we turned and went back to the entrance.

My guess is that we had explored some 3,000 to 4,000 feet of passage in the 7½ hours we spent there. There is a great deal more cave here than what we saw[1]

Figure 17-1. Walter Scheffrahn and Al Goodrich negotiate the junk at the bottom of Woodard Sink. Photo by Tom Cussen, 1978.

Whereas the trip of May 14, 1978, had discovered the cave that lay beyond the siphon, this trip had discovered a new entrance that would allow exploration of the new cave without scuba gear. This new entrance was named the Roy Woodard Entrance in honor of a former landowner. R.J.

---

1 Walter Scheffrahn, "Memorial Day Week-end—Dunbar Cave Instead," *The Michiana Caver*, vol 5, no. 6, pp 71–72.

Figure 17-2. Walter Scheffran's feet in the tight spot in the Roy Woodard Entrance crawl.
Photo by Tom Cussen, 1978.

Figure 17-3. Walter Scheffrahn rappels the Roy Woodard entrance pit.
Photo by Tom Cussen, 1978.

Heltsley mapped and named Roy Woodard as a 60-foot-long cave in his 1965 masters thesis, *Bats and Caves of the Northwestern Highland Rim of Tennessee.*[2]

Roy Woodard was the owner of the sink when Heltsley did his biology work there. Frank Meador was the owner when the Northern Indiana Grotto came to the project in 1978.

Figure 17-4. Walter Scheffrahn (left) and Al Goodrich (right) at the base of a frozen waterfall formation. Photo by Tom Cussen, 1980.

2 Thesis on file in the Tennessee Room of the Austin Peay State University Library, Clarksville, Tennessee.

Chapter 18

Signature Rock

Once the May 14, 1978, trip through the sump had proven that the cave continued, and once the May 28–29, 1978, trip had discovered the Roy Woodard Entrance and made further exploration feasible, the next step in the thorough exploration of the cave was to begin surveying the cave beyond the sump. Cave surveying is a tedious, time-consuming task, but it is the only way to document the length and size of the cave passages in relation to each other and the surface. For mapping purposes, the explorers referred to the new discoveries as the Roy Woodard Extension.

On June 24, 1978, Jack Countryman, Dan McDowell, and Roger Cole explored approximately 800 feet upstream from the Junction Room to determine if there was much cave in the upstream direction. Troglobitic crayfish and the cavefish *Typhlichthys subterraneus* were observed in the river. On July 3, 1978, they began surveying in Roy Woodard. Jack Countryman, Walter Scheffrahan, and Dan McDowell, along with Dave Doolin, Mark Najmon, and Dave Collins surveyed from the Woodard Entrance to the Junction Room plus some short side passages near the entrance for a total of 1,590 feet.

Labor Day weekend found Jack Countryman, Walter Scheffrahn, Al Goodrich, and Dave Collins with Skip Alwine and Tom Cussen surveying downstream from the Junction Room to the divers' tie off point. Due to surveying difficulties through breakdowns and deep water sections, two back-to-back trips of 11 and 12½ hours were required to survey the passage. Sixty-five survey stations netted 3,460 feet. These 12-hour surveying trips became the standard average for future trips into the cave. Roy Woodard surveyors would consider less than eight hours a short trip, while 16-hour trips were not uncommon.[1]

On the weekend of October 28–29, 1978, the survey team mapped over 3,000 feet of passages. Most of this footage was achieved in a single side passage with no end in sight. Located in this new discovery were several large, impressive domes. This brought the total amount of passage mapped in the Roy Woodard Extension to over 8,000 feet and the total length of the Dunbar Cave System to nearly 5 miles.[2]

Dan McDowell wrote the following account of this trip:

In the early trips we made to the Roy Woodard Extension, the upstream section of Dunbar Cave near Clarksville, Tennessee, we were quite aware of the total absence of signs of past visitation into that

---

1 Dan McDowell, unpublished manuscript, about 1982.

2 Anonymous, "Dunbar Cave News," *The Michiana Caver*, vol 5, no. 11, p 105.

section of the system, yet a few of us were aware of a reported visit to Roy Woodard sometime around 1940. In fact, it was this very information that was instrumental in our initial search for an entrance into the system on the vast sinkhole plain to the north of Historic Dunbar Cave.

Finally, on the trip of October 28, 1978, we found the signatures of several persons in the passage off survey station A-13. What was really interesting about all this was that we had explored or surveyed around two miles of passage and had seen no signs of any past visitation other than by raccoons. Then we found three arrows scratched into a mud-covered breakdown block about 2 feet above the floor. A short distance back toward the main passage we found the signatures, the names of four persons scratched into a large breakdown slab tipped between the left wall and the

floor. Several of us had walked right by the signatures and stooped beneath the slab, but with eyes on the passage beyond. The slab shows signs of having been under water, and mud and debris are deposited on its sloping horizontal top. Debris rings are present on the vertical side containing the names. We had some trouble deciphering them, but subsequent findings revealed them to be: Charles B. Staton, R.H. Harris (Roy), Billy Watts, Walter Watts, and the date, 10-14-39.

Billy Watts' name was the key to finding out who the people were. Billy's last name appeared to be spelled "Wells" or "Wills," and Walter's last name was almost obliterated. Staton's name appeared to be "Safton" or "Saffton." The one really legible name was R.H. Harris, and the date was quite legible. I copied down the names as best I could and later went to the Clarks-

Figure 18-1. Signature Rock. Photo by Tom Cussen, 1979.

ville telephone directory where I looked up all the names I thought might be possibilities. There were four entries for "R.H. Harris," one was an old cave explorer, but not the right one.

On the chance that one of the names might be "Watts" I recalled a Mr Watts whom I had met in 1973. He had taken my brother-in-law and me to Roy Woodard Sink and told us about visiting there as a teenager. He had taken us to the bottom of the sink to show us the general area in which he'd found a crawlway entrance into the cave. He had also shown us a place on the rim of the sink where a friend had fallen into the sink and sustained a concussion, leaving him in a coma for two weeks.

Anyway, on the chance that "Billy Watts" might indeed be Mr Watts, I visited him. It turned out that this was his name, and he had left it there on the trip he'd told me about in 1973. He now provided the details of a most unusual caving trip. I might add that this man's memory is phenomenal. Over 39 years later his description of the cave is highly accurate. The following account is of the trip as Billy Watts recently related it to me:

"... there was four of us, Charles Staton, Roy Harris, my younger brother Walter, and myself. It was my sixteenth birthday, October 14 [1939]. We were all about the same age except for Walter. He was fourteen. It was a real nice day. We went in about noon and we were in the cave for about four hours ... . None of us had ever been in a cave before. We took one flashlight and had several candles and a section of rope. We also took an old cap and ball pistol and a .22 rifle because we didn't know what we would come across in there. I remember I shot a lizard [salamander?] with the rifle.

"In the bottom of the sink, against the back wall, there is a little room. On the floor we went through a crawl that was low and we had to lay on our stomachs to get through. After the crawl we came out on the side of a room that had a 3- or 4-foot drop to the floor. A little ways from this room was a pit about 30 feet deep.

"After we tied off the rope it liked about five feet touching the bottom. When we come back to the rope to climb back out we had to pile rocks up against the bottom of the pit so as to be able to get a good high grip on the rope so we could use our feet to help us climb. We had a real hard time getting back up the rope because it was wet. It was a good thing that we were all in good shape.

"We went down the passage at the bottom of the pit. Where the passage made a bend we went between two large rocks. This passage was really muddy. I don't remember how far we went, but I guess it was several hundred feet. We turned back when the passage was sloping down to water along one side of the wall of the cave. We were afraid we might fall into a deep pool of water and might not be able to get out, or worse yet, go underneath a wall. On the way back out, we signed our names on a large rock that was slanted between the wall and the floor. We seen a few formations in one place in the cave.

"The reason we went into the cave was we thought there might be some buried money in there. On a nearby tree we found some old initials caved 'J.J.' and naturally we thought they referred to Jesse James. We found out later they were the initials of another person. Now that I look back at it, I guess it was pretty foolish, but we were young and it was an adventurous thing to do. Looking back, I think we were lucky that we didn't get hurt because of our poor lights and having to climb that rope and all."

I see two reasons why this trip of 1939 is deserving of attention. First is the earlier mentioned absence of signs of visitation in any section of the cave (with two minor exceptions.) The second is the character of the cave. Most people will not go through a crawl such as Roy Woodard has, then at-

Figure 18-2. Tom Cussen by a 10-foot-tall totem pole. Photo by Tom Cussen, 1980.

tempt to negotiate a pit, particularly if they have never been in a wild cave before. That this group made it as far as they did is an achievement in and of itself.[3]

On Monday, October 30, 1978, cavers Walter Scheffrahn, Jack Countryman, Dan McDowell, and Dave Doolin (representing the Northern Indiana Grotto) met with representatives of the Tennessee Department of Conservation in Nashville to discuss the future of the project at Dunbar Cave. As a result of that meeting, the Northern Indiana Grotto received thanks for their work thus far and were reassured of future cooperation and assistance.

Dan McDowell wrote the following report of the next major surveying trip:

On the trip of November 24 to Dunbar Cave near Clarksville, Tennessee, we discussed what sections of the cave to map.

We had about 30 known plotted leads on the map at the time. I wanted to map downstream because there had been only two trips into this section of the Roy Woodard Extension, and there had been no mapping in this area since the trip of Labor Day weekend. Also, I had not been in the downstream section and I wanted to see it. After some discussion, a survey team was formed consisting of Skip Alwine, Dave Doolin, Alan Goodrich, and me. Al and Skip had been in the downstream section, Dave and I had not.

Friday evening at 7:30 we entered the Roy Woodard Sink. We made the crawl, negotiated the pit bypass, and headed for the River Styx. We stopped briefly at the side passage at Survey Station A-13 to show the passage that contains the domes and the 1939 signatures. After this we proceeded to the river, turned right, and headed downstream.

The passage ranges from 8 to 25 feet high and the water is from a few inches to 3 feet deep. There were a number of blind

3   Dan McDowell, "An Early Trip to Roy Woodard Cave," *The Michiana Caver*, vol 6, no. 2, pp 13–14.

and surface crayfish in pools of water along the passage. After some 700 or 800 feet is Survey Station 45. At this point there is an upper level passage on the left that Al had remembered seeing.

When we first started surveying into the passage from the river, I was not very impressed. But as the survey progressed things began to improve. First, all the stations were generally 70- or 80-foot shots in a large passage measuring 5 to 20 feet high and up to 30 feet wide. A small stream was present and the entire floor was muddy throughout the passage.

Then we came to a nice display of calcite crystals up to 2 inches thick with twined intergrowths. Immediately beyond this area, at Station 45.11, the passage split into two major passages. We took the left passage, which ran almost perfectly straight for about 450 feet to a large breakdown room (later named the Domino Room). About 200 feet long, 60 feet wide, and 30

to 35 feet high, the room contained four leads we plotted. We surveyed over the top of the breakdown and through the middle of the room. Beyond, the passage changed into a canyon-like passage about 200 feet long that contained some attractive formations. The passage then dropped to about 2 feet high, and we stopped the survey at this point.

Retracing our steps back to Station 45.11, we proceeded to map into the right hand lead. For the first 17 stations there was not much of interest and the passage was becoming rather small, some stoopway and short crawlway sections.

Then suddenly, the ceiling rose up to 30 feet in a large, spacious room with several scattered stalagmites, the largest being 6 feet high and 3 feet in diameter. The dimensions of the room are 40 by 60 feet with the floor rising to the ceiling at the back of the room (later named the Paradise Room). In a smaller alcove on the right

Figure 18-3. Tom Cussen admires beautiful flowstone formations. Photo by Tom Cussen, 1980.

Figure 18-4. Beautiful cave formations with macrocrystalline flowstone. Photo by Tom Cussen, 1977.

side is a well-decorated section containing beautiful stalagmites that are pure white. Some of the formations are red and orange. This is by far the most attractive section we have seen in the Roy Woodard Extension of Dunbar Cave.

We terminated our survey on the left wall at the back of the main room where we will be able to make the next survey shot into the continuation of the passage. It will go a short distance to another formation room. About 5 feet from our last station was a solitary hibernating bat.

We arrived at the surface at 8:00 Saturday morning, having been underground for 12½ hours. In that time we had surveyed about 2,100 feet of virgin cave in 33 stations. We had discovered a breakdown room almost as large as the Stone Mountain Room in the Historic Section of Dunbar Cave, and had found a very attractive formation room. In addition, we had noted or plotted several other leads, some of them to walking passages. It was a very

rewarding and productive survey trip. The Roy Woodard section of the cave system is proving to be quite extensive.[4]

That same weekend, a second survey team was also in the cave. Skip Alwine describes their efforts:

Saturday evening, November 25, Mark Najmon, Alan Goodrich, Dave Doolin, and I entered Roy Woodard Cave to survey and explore, our main goal being the passage upstream from the Junction Room. On a previous trip, Jack Countryman and Dan McDowell had partially explored the upstream section. We intended to penetrate as far as they had, and then beyond.

After an uneventful trip through the entrance crawl we descended the 30-foot pit and continued on to the Junction

4   Dan McDowell, "Dunbar Cave—New Discoveries in Roy Woodard," *The Michiana Caver*, vol 5, no. 12, pp 113–114

Room. Here, we began our survey.

The trip required full wet suits. Upstream, the passage is low. In places, the water is inches from the ceiling. Everything is coated with a thick layer of mud, reminiscent of axle grease. We duck-walked and crawled through this for about 1,000 feet, finally arriving in a small room where we stopped and rested.

To the left there is a small hole in the wall, about 18 inches from the floor. It is just large enough for one's body to pass through. The continuation of the passage is through this orifice. Low ceilings and wet, gravely floor tormented us for about 200 feet, after which we climbed up into a breakdown room.

Loose rock congested the entire floor up to within a few feet of the ceiling. To all appearances, the passage seemed to terminate. While Mark and Dave were checking for other leads, Al and I decided to check a small opening in the floor near the wall. It was wet going, but soon we were able to walk.

After a short distance, the passage seemed to terminate in another mass of breakdown forming a plug and barring any further advancement. Yet overhead we could see what appeared to be a possible continuation if we were willing to climb over the breakdown. We threaded our way up, towards this inviting darkness that loomed just beyond our reach. Without warning, we found ourselves in a huge, open room. Its floor was covered with a mountain of limestone blocks. Combined with the enormity of the room, the massive blocks made us feel small and insignificant.

Off in the distance the roar of a waterfall demanded further investigation. In the dim light of our carbide lamps, we were

Figure 18-5. Large stalagmite in Independence Hall with a rimstone pool at it's base.
Photo by Tom Cussen, 1977.

able to see the ghostly form of a waterfall as it plummeted over the edge of a precipice into a deep basin.

With the renewed aid of Mark and Dave, we began searching for some means of getting to the top of the waterfall. We finally located a ledge that snaked its way towards the balcony from which the waterfall issued.

Behind the waterfall, upper-level passage meandered off into the darkness. Its ceiling and walls were generally uniform in contour, with 10 to 15 feet of ceiling height and 30 feet of width being the average dimensions of the passage. The stream had a tendency to favor one side, occasionally disappearing beneath breakdown, to reappear around the next bend. We pushed forward until fatigue became a problem.

After a short rest, we decided it was time to head out.

We are quite confident that we were in new cave. The fragile beauty of its passages continued on, but our own physical limitations dictated that we turn back.

Slowly we retraced our steps back to where we had left the survey equipment. We collected it and our personal gear, and reluctantly headed towards the entrance.[5]

On every trip into the cave the explorers were finding fascinating new sections of the cave and adding significantly to both its surveyed and explored length. By the end of 1978, 8,100 feet had been surveyed in Roy Woodard Cave. Dan Mc Dowell now realized that the reality of the Roy Woodard Extension becoming a substantial find was starting to come true.

---

5   Skip Alwine, "Upstream in Roy Woodard," *The Michiana Caver*, vol 5, no. 12, pp 114–115.

Chapter 19

Watts Trail

Exploration and surveying continued at a steady pace in 1979. Dave Doolin gave the following description of the March 31, 1979, survey trip:

A little more than 1,100 feet were surveyed in the upstream continuation from the near-sump in the Roy Woodard Extension of the Dunbar Cave System during the March 31 trip. Members of the team were Skip Alwine, Al Goodrich, Dan McDowell, Mark Najmon, and me. The passage past the near-siphon turned into a breakdown crawl, which opened into a large room containing a waterfall of large volume and walking-size stream passage leading in a northerly direction. Several hundred feet upstream, a dry overflow route led into a fairly large room (later named the Acheron Room) approximately 200 feet by 80 feet by 30 feet high, with the stream flowing along the eastern edge. A few hundred feet further on an immense breakdown plug seals the passage. The water emerges from beneath it and, because the water was up a foot, we couldn't find a way through. A possible overflow route was found, but it will require some digging. There appears to be no way through or over the plug. It is incredibly unstable, with loose rock projecting at every angle from the floor and walls. I would consider it quite easily the most dangerous part of the not-so-small portion of the cave I have yet seen in this segment of the system. Since there are only a couple of really good leads in the segment between the waterfall and the collapse, this area should be fairly easy to wall off and complete, thus allowing greater concentration on other areas.[1]

The next survey trip to the cave occurred on the weekend of July 6, 1979. Erika Berger describes the trip as follows:

The evening of July 6, Skip Alwine arrived alone in Fort Wayne where Walter Scheffrahn and I waited. It seemed that the trip to Dunbar was going to be made by just the three of us since some grotto members stated earlier that they'd be unable to attend the scheduled trip. We hoped to survey as much as possible with our small group. We were pleasantly surprised to find Dave Doolin asleep in his car when we arrived at Dunbar. That wasn't the only surprise. We practically drove right by the bathhouse without recognizing it because of the renovation of the old structure. There is now a new roof and exterior walls. The windows have not yet been installed and work is still needed on the interior walls. Plans for the swimming pool now call for an elegant flower garden with brick pathways throughout. The intended ramp from the bathhouse to the cave for the dis-

---

1   Dave Doolin, "Roy Woodard Cave: March 31," *The Michiana Caver*, vol 6, no. 5, p 44.

abled was not yet under construction. All the holes from the past excavations in the cave entrance have now been filled in with concrete. Soon, work will begin on the old concession stand. Paul Huff met with us on Saturday morning to update us on the development of the natural area.

Afterwards, we went to breakfast at the nearest Pancake House. While there, we gave Ed Vengrouskie a call to inform him of our plans at Roy Woodard. He promised to join us. Then we went out to the cave.

We chatted for a time with Mr Meador. I was glad to meet Mr and Mrs Meador and was most thankful for their kind hospitality. By this time, Ed had arrived with his brother, Fred, who had come along to help him collect insect specimens.

It started raining as we descended the junk-filled sinkhole. I must admit that I was rather apprehensive climbing over, around, and through those dilapidated cars. I was finding out first hand why they wanted to clean up the area in order to cut

down on the risk factor. But that's not all I learned. I had the thrill of experiencing the famous belly crawl for getting into the cave. Talk about a tight fit. Lucky for me that I'm not a big person. Walter had the privilege of being the first person to enter. His was thus the task of initially cleaning out the perpetually filling crawl for the rest of us. We finally got through without incident.

The rope was still there from the previous trips, which made it a bit easier to make the 35-foot drop. From here we headed for the dry passages beyond the Signature Rock. Dry? The way was muddy and sloppy. Several times Ed stopped to pick up bugs that we found along the way to virgin passageways. We found some cave flowers, unbelievably long, and numerous soda straws. We discovered a new dome room. While the others surveyed, I went ahead to poke around on my own. I felt a sense of anticipation being the very first person to walk upon these muddy floors, which,

Figure 19-1. Al Goodrich in the River Styx at the Junction Room. Photo by Tom Cussen, 1978.

Figure 19-2. Large passage in the Woodard Section. Shown left to right: Jack Countyrman, Walter Scheffrahan, and Pat Harvey, along with two unidentified cavers. Photo by Tom Cussen, 1978.

to me, represented the desert windswept sands. I had a brief imagining of what the men walking on the moon must have felt like.

By now we had spent seven hours in the cave. Half of us wanted to leave while the rest wanted to stay. I, for one, especially wanted to get out, I had learned a valuable lesson, my system does not tolerate drinking cave water. Because of this situation, the others agreed to leave.

The journey back seemed to be the longest, most miserable trip ever. When we finally did reach the surface I was really exhausted. Even if I hadn't had the problem I would still have been tired.

The night passed quickly and morning found us awakening to a steady rainfall. After some discussion, we decided to not go back into the cave, but rather head on home. Looking back now, it seems humorous, but at the time it seemed like a forlorn

weekend. Still, in all, it was a good trip.[2]

That same weekend, Dave Doolin describes the actual surveying activity that occurred:

Walter, Skip, and I surveyed while Erika, Ed, and Fred followed along doing whatever biology people do. The area we decided to work was the Up & Down Series and the area in the A-13 discoveries. We found some fairly nice passage beyond the previous end-of-survey point and managed to survey a slow 850 feet from there. The passage looks much like the previous, requiring constant climbing up and down over mud banks but staying generally at about the same level as a massive bed of chert nodules. A small stream ducked in and out for a few feet, and we found a classic set of superb domes. The highest dome

2 Erika Berger, "Forlorn weekend ... at the time," *The Michiana Caver*, vol 6, no. 9, pp 71–72.

has a hole in the center of its top and it is probably 70 feet high. Along with all this was the usual assortment of good and bad leads. This area will almost certainly yield a mile or more of cave. It shows no signs of stopping.[3]

According to an Editor's note at the end of Dave Doolin's article, this survey trip brought the total length of mapped passages to 6.3 miles. This total included both the Historic Section and the Roy Woodard Extension. The "siphon" connecting the two sections still remained to be surveyed. Exploration and surveying seemed to be winding down. Dan McDowell summed up the final trips of 1979:

> Although there has been a slump in Northern Indiana Grotto activity at Dunbar Cave over the last several months, there have been some things going on. Last October 27th (1979), Jack Countryman, Ramon Costello, Steve Clarke, and I made a 9-hour exploring trip into Roy Woodard. Most of the time was spent in that section of the cave we have named Watts Trail. We discovered three new side leads. One goes about 200 feet off the main passage into a really spectacular dome. About 80 feet high, it is 20 by 30 feet in size, top to bottom. This one is the prettiest dome of the several we have found in Roy Woodard. About 30 feet above the floor is a passage 10 feet wide by 15 feet high leading off somewhere.
>
> Some 300 to 400 feet down Watts Trail, Steve climbed a difficult ascent that opened up into what may be an extensive upper level. He checked out about 300 feet

of walking passage that crossed over the top of Watts Trail.

> A third lead opens up in a set of double domes right off the main passage. The domes may be known to the team who surveyed Watts Trail, I don't know, but these double domes are not on the latest edition of the map.
>
> On the Friday after Thanksgiving, Skip and Tom Alwine and I went into historic Dunbar Cave for a leisurely 5-hour trip. An interesting thing happened. We went across the River Styx and into Indian Trail. We were there about an hour. When we came back the river's water level had risen a foot or so and covered the bridge across the river. There was a lot of rain that weekend. I went back into the cave on Sunday afternoon about 2:00 P.M. to check the water level. The river was out of its banks in the Entrance Room. It looked like a small lake. If the water had risen another 5 or 6 inches it would have started to cover the passage leading to the main part of the cave.
>
> An interesting thing happened last October. Land was being cleared directly across the road from the lake. In heavy underbrush, leaning against a tree, the tombstone of Isaac Rowe Peterson was found. Peterson was the first legal owner of Dunbar Cave. The stone measures 46 inches high, 18 inches wide, and 3 inches thick. The inscription reads: "In memory of Isaac Peterson deceased July 22nd, 1833, aged 78 years."[4]

This is the stuff cavers' dreams are made of: finding, exploring, and surveying thousands of feet of virgin cave passage.

3   Dave Doolin, "Roy Woodard ... finally," *The Michiana Caver*, vol 6, no. 9, pp 72–73.

4   Dan McDowell, "1979 Wrap-up at Dunbar Cave," *The Michiana Caver*, vol 7, no. 2, pp 12–13.

Chapter 20

The Domino Room

The Northern Indiana Grotto planned a trip for February 9, 1980, to map the siphon and "officially" connect Dunbar Cave to Roy Woodard Cave. Unfortunately, this trip had to be cancelled when one of the divers backed out. However, on February 16, a total of 23 people from five different Grottos showed up for a tour of Roy Woodard Cave. Two Nashville Grotto members, David Parr and Joe Douglas, were given a tour of the Roy Woodard Section of the cave, so that if the Nashville Grotto's Rescue Team were ever needed, that they would already be familiar with some of the cave's main passages.

Another trip occurred on February 16, with members of the Bloomington Indiana Grotto, Northern Indiana Grotto, Western Indiana Grotto, and Nashville Grotto present. The ten people present split into two teams of five each. Quite a few climbs and difficult leads were checked, but no surveying was done on this trip.

A trip on July 12, 1980, by Jack Countryman, Ray Costello, Gary Doolittle, Bill Svihla, Keith Dunlap, Mark Pies, and Dave Pierce went into the cave through the Woodard Entrance and back to the Twin Waterfall Domes at the end of Watts Trail. The group succeeded in using a scaling pole, designed by Keith Dunlap, to reach the top of one of the domes. Due to the time involved in carrying the scaling pole into the cave, little exploration of new passage was accomplished that day.[1]

The next survey trip took place on the weekend of August 9, 1980. The trip report is as follows:

At 11:00 A.M., Saturday, we entered Woodard and found the entrance has collapsed as reported on the July trip report by Jack Countryman. Because of the collapse, getting through the crawlway seemed to take longer than usual.

With Walter taking notes, Dave reading compass, and I reading the tape, we planned to survey as much footage as possible. We had a two-fold strategy: First, survey the leads in the Domino Room and the Paradise Room with all its leads. Second, push and survey up the Rimstone River and its side passages.

We found that the water level was low as we waded down the river to the Gumbo Avenue. (Later renamed Grand Avenue.)

From there we headed to the Domino Room where we measured the room to be more than 100 feet wide. We surveyed to a large 20-foot pit and what appears to be a canyon passage beyond. The unusual feature about the pit was that it consists entirely of clay. Beyond the Domino Room, we pushed one of the leads to its end.

---

1 Jack Countryman, "Roy Woodard Cave – Trip Report," *The Michiana Caver*, vol 7, no. 7, pp 61–62.

Figure 20-1. Gary Doolittle climbs the scaling pole.
Photo by Dave Pierce, 1980.

squeezeway.

On our way back to the River Styx, we stopped to eat and change carbide. There we noticed a passageway entrance that had an appearance similar to that of a window and door of Fred Flintstone's house. We surveyed into that passage that lead to a moderate-size dome room. The room had three leads which all eventually pinched off.

Then we proceeded to the Frozen Waterfall. Seeing it for the first time made me realize that the pictures I've seen didn't really do it justice. It wasn't as difficult as I thought it would be to climb the waterfall. Once on top, we began to survey the Rimstone River Passage. I was overwhelmed by the size of all those rimstone dams. We surveyed some 300 feet up the river passage. After noting two leads, we discovered and surveyed a third lead which headed toward the River Styx.

It was approaching a 14-hour stay in the cave, my longest record so far. We all agreed to head out where the heat of the night air greeted us at the entrance. It was around 1:30 A.M. when we were searching

Backtracking, we headed towards the Paradise Room to survey the well-decorated White Place and its small dome and short side leads. Another room off the Paradise Room is the Hidden Paradise, which is even more beautiful and fragile than the White Place. To get into it to survey, we had to contort our bodies up a tight

for a place to eat in Clarksville. Luckily there was still a place open. It certainly felt good to have a warm meal in my stomach. Soon they closed, so we trecked back to our campsite at the Woodard Entrance for a well deserved rest.

Sunday morning we ate breakfast, packed, and without much delay, headed for home.[2]

Three weeks later, the Northern Indiana Grotto cavers were back, hard at work on their survey. Gary Najmon describes their trip:

Late Sunday morning, August 31, 1980, Erika Scheffrahn, Kathy Goergen, Mark and Ralph Najmon, and I went into Woodard Cave on our way back to the Domino Room. Erika, Kathy, and Mark were to survey the passage at the end of the Domino Room and Ralph and I were to explore the area for new passages.

Figure 20-2. Walter Scheffraham and an unidentified caver in Hidden Paradise. Photo by Tom Cussen., 1979.

2   Erika Berger, "Dunbar Cave System Surpasses 10 Kilometers," *The Michiana Caver*, vol 7, no. 8, pp 65, 73.

Erika suggested that we start by climbing down a hole in the breakdown. A large piece of loose rock and much loose dirt made our safety doubtful there, though it seemed to go rather far. Erika then suggested a passage at the far end of the Domino Room just before it turns to passage again.

On the left there is a 18- to 24-inch-high, 15- to 20-foot-wide passage.

It didn't look like it went, so I told Ralph to check it out. Soon he yelled back, "There's a trench in here and it looks like it goes a little way ... Hey! It's getting bigger, I can stand up and it goes!" Well, by that time I realized I was missing something so I crawled in, carefully avoiding the numerous formations. There was the trench, just starting out in the middle of the floor, seemingly without reason for its position. Soon I had caught up with Ralph.

As we went on, the passage was fairly uniform, about 2 feet high, 15 feet wide, but the height available was reduced by about 8 inches because of the many soda straws covering every inch of the ceiling, necessitating crawling in the trench.

The passage seemed to be a large walking passage which had been partially filled with mud and uniformly smoothed out. Then natural seepage caused the forma- tion of both the straws and the trench. As we progressed, the trench got bigger, and we came upon the first of several rooms, which were really just areas down into the mud with the ceiling at the same level over the rooms as in the crawlway. To proceed, we climbed up out of each room onto the high mud floor.

Some of the rooms looked as though they had good possibilities for side passages, but lack of time decreed that we couldn't check them out. One looked very wet and tight. There were also two or three places that looked as though a little climbing might lead to some upper-level passages. Another place looked as though it was very close to the surface. A hole went up into the rock and the walls appeared to be loosely supported. A little digging might create another entrance. At that point it seemed as if large quantities of rainwater had funneled in to wash away the mud. There was a pile of fist-sized rocks

Figure 20-3. Cavers wade in the River Styx. Photo by Tom Cussen, 1977.

Figure 20-4. Dan McDowell (left) and Don Nehring (right) at the base of a breakdown slope. Photo by Tom Cussen, 1977.

expect from curious cavers?)

Finally, "Ralph, let's leave our stuff here and go around the bend. That way we'll be forced to turn around soon." Well, just around that bend was a short crawl, which ended in a small room with a high ceiling. This seemed to be the end of the passage, so at last there was no objection to turning around. While we stood in the room, we began to wonder just how far we were from the Domino Room. A brief discussion ended in the profound conclusion that we didn't have the foggiest idea. We decided to estimate the distance on the way back by both of us counting off the feet and keeping independent totals as we went out.

As we approached the Domino Room, we heard Mark yelling at us. The others had finished surveying and Mark had come in to look for us. When back with the others, we compared estimates of the distance we had explored. My guess was about 1,000 feet and Ralph's was about 1,150 feet.

A considerable portion of the ceiling of the passage had been covered in soda

and loose, sandy dirt that appeared to have fallen from the hole.

After a while I said, "Ralph, we've got to turn around soon to be back when we're supposed to be."

"Just a little farther," was the reply.

I repeated myself often, but we just kept going. (Well, really, what would you

straws, and with the estimated length and width of the passage, a rough calculation puts the number of straws in the passage at about 100,000. I do not believe that this is an exaggerated number, as I have never seen more in one place before, and they continue down the passage consistently.

At this time, we headed out of the cave, ending one very enjoyable cave trip with a new find and immediate hopes of much more virgin passage.[3]

What a trip! Ralph and Gary had discovered and explored over 1,000 feet of virgin passage, which was filled with 100,000 soda straws. That same weekend another survey team of Walter Scheffrahn and Dave Doolin was in the cave. Their goal was to finish the loop survey beyond Billy Watts' signature in Watts Trail and to survey a smaller loop in Frank Meador Passage.[4]

On the weekend of October 24 another group of Northern Indiana Grotto and Western Indiana Grotto members entered through the Woodard Entrance to survey Upper Watts Trail. This trip resulted in approximately 1,000 feet of passage being surveyed in 81 survey stations.[5]

On the weekend of November 27, 1980, which was the Thanksgiving Holiday weekend, Keith Dunlap, Bill Svihla, and Dan Solomon decided they would attempt to push to the end of the Northwest Passage in the Historic Section of Dunbar Cave. This push trip resulted in the discovery of over 1,000 feet of virgin passage. The next day, the explorers returned with Walter Scheffrahn and mapped their new discovery, which they named the Southeast Discovery. Meanwhile, Dan McDowell, Erika Scheffrahn, Gerry Grieve, and Mike Sunderman surveyed over 1,000 feet of passage at the north end of the Domino Room in the Roy Woodard Section of the cave. Therefore, approximately 2,100 feet of surveyed passage was added to the map.

By the end of 1980, several different grottos were now involved in the Dunbar Cave Project and considerable amounts of new cave were still being discovered and mapped. Dunbar Cave was now established as a major cave, not only in Tennessee, but nationwide. The August 1980 issue of the *NSS News* published a list of the 100 longest caves in the United States. Dunbar Cave ranked 77 on the list at a length of 5.250 miles. And this listed length was from an April, 1979, report, nearly 16 months out-of-date.[6]

---

3   Gary Najmon, "Woodard Cave Trip Report: A New Find!" *The Michiana Caver*, vol 7, no. 9, pp 75, 85.

4   Walter Scheffrahn, "Pushing Deeper in Dunbar," *The Michiana Caver*, vol 7, no. 9, p 78.

5   Bill Svihla, "Upper Watts Trail," *The Michiana Caver*, vol 7, no. 11, pp 102, 104–105.

---

6   Bob Gulden, "The Hundred Longest," *NSS News*, vol 38, no. 8, pp 186–187.

Chapter 21

Portal to Portal—Dunbar to Woodard

The long-awaited trip to survey the water-filled passage between the Historic Section of Dunbar Cave and the Roy Woodard Section of the cave finally came on the weekend of January 3–5, 1981. Groups from both the Northern Indiana Grotto and Michigan Interlakes Grotto converged on Dunbar Cave for the attempt. Plotting the survey data from the last trips in 1980 had shown that there was also a possible connection at the end of the Northwest Passage. The map showed only a 200-foot separation between the two sections of the cave.

Dale J. Purchase describes the effort by the divers:

On Saturday morning we planned the dive to survey the River Styx sump. Walter Scheffrahn, Gerry Grieve, and Keith Dunlap moved out to meet the dive team at the Woodard end of the sump. Another group helped us gear up, and a third group headed out to explore the Dunbar Northwest Passage to try to make a dry connection. Just before the dive, Fred LaClair had a series of problems, which led to his missing the dive. First his wet suit zipper broke, then a fin was lost in the black water, and finally his light burned out. Fred decided that this was not his day and wisely sent the three of us on without him. Bruce Herr led the way and laid the safety line. John Garner followed and took the compass heading, and I brought up the rear with one end of the survey tape. We covered 576 feet, 350 feet of which was totally under water. Bruce had about 15 or 20 feet of visibility, but due to the small passage and a large amount of silt, the visibility was less than a foot for me. The water survey took us about one hour. Then we put our gear on the riverbank and moved on up stream about 300 yards to wait for the party from the Woodard end. About an hour later they found us.

We ate lunch and then we all went back down stream to near the point where the sump begins. I have to give a lot of credit to Walter, Keith, and Gerry, who followed us through water, where they could not touch bottom, with only 8 to 14 inches of air space over the water. This situation can be a bit uncomfortable if you are not used to it. We found a large room and a small side passage. John and Bruce then moved on back through the sump in zero visibility. John had to take my tanks and fins back with him. Bruce took my light pack and handled the safety line.

That was not an easy dive. Picture yourself dragging a 90-pound set of double tanks, an extra light, and extra fins through a small tunnel following a 3/32-inch line in zero visibility. The trip took them 27 minutes, all underwater.

The rest of us moved on toward the Woodard Entrance. The trip out was a combination of wading in chin deep water

Figure 21-1. Divers prepare to enter the sump for the survey. Dale Purchase on the left and John Garner on the right. Photo courtesy of Tom Cussens, 1981.

(we took a new, lower passage that has not been surveyed), climbing through breakdown, slipping through crawlways, and climbing vertical passages. Roy Woodard Cave is filled with many beautiful formations: flowstone, waterfalls, rimstone dams, soda straws, bacon rind, and almost every other type of cave formation you can imagine. It was a beautiful, but strenuous, trip. It took us about 3 hours and 45 minutes to get out. My total trip lasted eight hours from the start of the dive to the top of Woodard Sink.

We then moved on through the tight part of the crawlway, out past the junked cars and into the cold, dark night. It was a big drop in temperature to go from a 60-degree cave to 27 degrees outside. We went back to Dunbar and on out to eat about 9:30 that night.

It took a big operation and the help of a lot of people from the Northern Indiana Grotto to get one person through from portal to portal, but I believe that it was all worth it. Thanks again to those people who carried our dive gear, guided me out of Woodard, and made this a successful trip. I hope to join you again.[1]

Dale Purchase became the first person to make the portal to portal trip from the Dunbar Entrance to the Woodard Entrance. And after nearly four years of surveying, the Dunbar Cave survey and the Roy Woodard Cave survey could finally be combined into one, large, master map.

---

1   Dale J. Purchase, "Portal To Portal – Dunbar To Woodard," *The Michiana Caver*, vol 8, no. 2, pp 17–18, 16.

Chapter 22

So Many Caves, So Little Time!

After four years of exploration and mapping, Dan McDowell summed up their project:

At the beginning of the project we, the grotto in general and I in particular, vastly underestimated the probable extent of the cave. From several years of local caving I figured the length of the cave would extend some 8,000 or 9,000 feet and possibly 10,000 feet maximum. We now have over four times the original estimates mapped. We did not foresee the cave turning into a system with an extension larger than the old Historic Dunbar Cave.

The two different sections of cave, the Dunbar Section and the Roy Woodard Extension, provide a sharp contrast to each other in appearances. Dunbar is basically a single level cave with extensive amounts of fill present throughout much of the south and east portions of the cave. This fill is of various depths up to 40 feet in some places. The variances in the fill present do provide different levels in the cave but these are not true developed levels as the term is usually understood. As a whole it is relatively dry and for all intents and purposes there is only one stream present, the River Styx. There are many interconnected passages and this section takes on maze-like characteristics. Also due to the numerous old commercial trails there is probably a tendency to somewhat misjudge the overall character of the Dunbar Section. In the extreme east, north, and northwest passages of Dunbar the cave appearance is more akin to the Roy Woodard Extension that lies to the north.

The Roy Woodard Extension is a true multi-level cave. The River Styx is the base level for over a mile of stream passage. This stream has downcut its passage leaving the Domino Room Passage at a level some 10 feet above the present stream. Watts Trail is not connected onto the River Styx, but it is at a corresponding level of the Domino Room Passage. Sporadically along Watts Trail, there are upper levels present and the Upper Watts Trail is a true upper level with its own stream. Even without the vertical survey data to back it up there is at least a 30-foot difference between the upper level of Upper Watts Trail and the River Styx and this difference may be as much as 45 feet.

The Roy Woodard Extension also has some smaller tributary or auxiliary streams besides the main River Styx. The Rimstone River Passage is a sizable tributary that comes in from the west. There is a small developed stream in the Domino Room Passage. The stream in Upper Watts Trail drops into a pool at the base of the Discovery Domes. There is another small stream in the Frank Meador Passage; we don't know its source or its ultimate destination. Roy Woodard Cave seems to be taking water from different areas of the overlying sinkhole plain and feeding it into the River Styx and from there into the Dunbar Section and then out into the lake.

With the exception of some small loops off and back onto a few passages there is none of the circle route or maze development in Roy Woodard that is present in the Dunbar Section. The Woodard map looks like a spider's legs heading off in all directions. At the present there appears there may be only a very limited possibility of any of the passages connecting back into one another.[1]

There comes a point in any survey and exploration of a truly major cave where motivation and manpower began to wane. As Dan McDowell noted:

The whole membership of the Northern Indiana Grotto does not participate in the Dunbar project. There is a core group of around ten or twelve people in the grotto who are consistently active in the project. This is not too hard to understand since we are a grotto located in northern Indiana and an average trip distance for the grotto is over 800 miles round trip. We seem to have settled down to some four or five trips a year to Tennessee and even then not everybody makes every trip due to various reasons.[2]

Clearly, by 1981, exploration and surveying was beginning to wind down in the Dunbar Cave System. Furthermore, the Northern Indiana Grotto was very active in the exploration of other caves both in Tennessee and Indiana. So many caves, so little time! Although the 1982 issues of *The Michiana Caver* mention several exploration and sightseeing trips into the Dunbar Cave System, no mention of surveying is made.

Due to increasing trash in the Woodard Cave Section by locals, the prospect of gating the Woodard Entrance was raised. As Keith Dunlap pointed out:

What has been observed so far is just minor trash that can easily be kept under control with an occasional cleanup. However, irresponsibility can also lead to carbide dumping, graffiti, and ultimately, damage to irreplaceable formations. A greater danger could be the possibility of an injury (especially if the cavers are under the influence of alcohol). Rescue and evacuation through the Woodard Entrance Crawl would be a very serious problem. Any incident would probably limit access for all and damage our favorable landowner relationship that we have with Frank Meador.[3]

In an August 27, 1983, trip report, Clyde Simerman noted the following careless damage to formations:

We spent a little time taking photos in the Hidden Paradise Room and then went on to the Domino Room and the Broken Domes. We were saddened to note the mud tracked across pure white flowstone floors and muddy handprints on pure white stalagmites. While the first may clear themselves, the latter need to be washed off.[4]

Although the 1983–1984 issues of *The Michiana Caver* reports numerous trips to the Dunbar Cave System, there is no mention of any active surveying during those years. However, Keith Dunlap describes a trip on July 20, 1985, where surveying continued:

---

1 Dan McDowell, "The Dunbar System: Reflections, Observations, and Speculations," *The Michiana Caver*, vol 8, no. 3, pp 23, 26–31.
2 *Ibid.*

3 Keith Dunlap, "Conservation—What is Needed in Woodard," *The Michiana Caver*, vol 10, no. 1, p 181.
4 Clyde Simerman, "Trip Report: Roy Woodard Cave, Clarksville, Tenn., August 27, 1983," *The Michiana Caver*, vol 10, no. 9, p 233.

The Northern Indiana Grotto weekend in Tennessee included several groups going into Woodard. One of these was to be a surveying trip into Upper Watts Trail. The survey team consisted of Gary Doolittle, Don Miller, Bill Sullivan (all from Terre Haute) and me. Upper Watts has not been visited for several years because of the 25-foot waterfall climb required to access it. This would be the third trip, following the original discovery trip and one other survey trip. Time has a way of dulling the mind because I remembered that the only passage remaining to be surveyed was an upper-upper level crawlway out of Surprise Dome paralleling the lower main passage. Thus this mop-up was my intention.

The trip back to the Twin Domes was normally an easy and enjoyable stroll. Unfortunately it was marred by evidence of recent visitation by major-league assholes who smashed the nicest formations along Watts. They had also marked their way with silver spray paint. When we reached Twin Domes, we were all fit to be tied and talked about aborting the trip and heading out. But after cooling down (literally, in the waterfall) we decided to continue. It took a while for us to rig the cable ladder. I then climbed the waterfall wearing a poncho and deflected the water for the others. Bill ascended and Gary went looking for Don who had wandered off. It seemed that Don was poking around back in the canyon leading to the Domes and had climbed up to a place he couldn't get down from. Gary finally "rescued" him and the two returned to the Domes and ascended to Upper Watts. It was apparent that Don was mentally shaken. The planned survey-

Figure 22-1. Muddy Cavers. Shown left to right: Walter Scheffrahn, Al Goodrich, and Mark Najmon. Photo by Tom Cussen, 1978.

ing was hampered because we could not locate any of the previous survey stations, so we had to start over at the top of the waterfall and resurvey up to Surprise Dome. The surveying was slow because of inexperience of the crew. The new footage consisted of a tight, muddy crawlway. The survey was terminated when Don's size exceeded the passage. We did pick up a little extra footage in a short virgin canyon leading in the opposite direction out of Surprise Dome, heading back towards and above the Twin Domes.

After packing up the survey gear, we headed towards the end of the cave for "sentimental reasons." Much to my surprise, I had forgotten how nice the passage was and how much cave existed. I had also forgotten that there was unsurveyed passage at the end of approximately 1,000 to 1,500 feet.

On the trip out, we "removed" most of the silver paint and I carried out the offending spray can (which decided to leak all over the inside of my cave pack). We again verbally maimed, dismembered, and mutilated the scum-sucking geeks as we passed their wanton destruction. We spec-

Figure 22-2. Gary Collins emerges from the gate to the Roy Woodard Entrance. Photo by Clyde Simmerman, 2003.

ulated that it's probably locals visiting the cave. Something needs to be done before they have a field day in Hidden Paradise. Then it will be too late.[5]

The weekend of August 31 to September 1, 1985, saw more mop-up surveying. Steve Reesman reports that on this trip they discovered the largest gypsum flowers that he had ever

---

5   Keith Dunlap, "Woodard Cave Trip Report," *The Michiana Caver*, vol 12, no. 9, p 59.

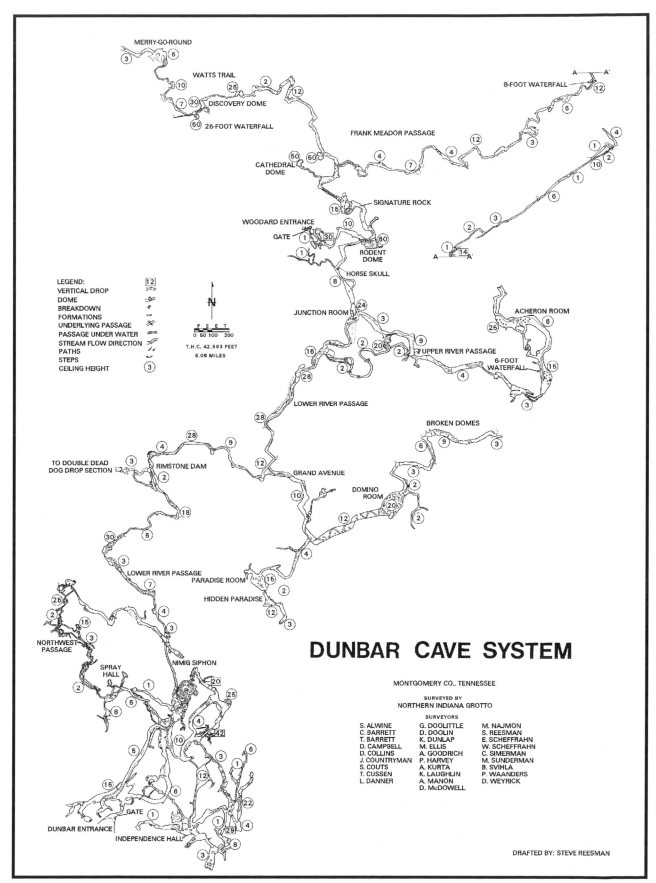

Figure 22-3. Map of the Dunbar Cave System, July, 1989.

seen in a cave.[6] The Northern Indiana Grotto's activity for 1985 was summed up as follows: "In 1985 seven survey teams made their way into the Woodard portion of the cave. They surveyed 3,379.5 feet of cave, most of which was crawlway, bringing the total cave footage to 42,593.7 feet (8.066 miles).[7] As any experienced cave surveyor will tell you, in any large cave system when it gets down to mapping all the side crawlways, motivation is difficult to sustain and volunteers are hard to come by.

In the October 1985 issue of *The Michiana Caver*, Keith Dunlap proposes that a gate be built on the Woodard Entrance to prevent any more vandalism in the cave. He notes, however, that he has proposed this several times before, but that there is still no gate. This time, accurate measurements of the site for the proposed gate are made and detailed construction plans are drawn up. Finally, on the weekend of March 28–30, 1986, a large crew of Northern Indiana Grotto cavers constructed a 2,100-pound cement and steel gate inside the Woodard Entrance.[8] Although some vandalism had occurred in the cave in recent years, this would prevent any future damage.

* * * * * * * * * *

Finally, the July, 1989, issue of *The Michiana Caver* announced that the completed map of Dunbar Cave was presented to Head Ranger Bob Wells at Dunbar Cave State Natural Area on May 20, 1989. That map has been framed and is currently in the lobby of the Visitor Center. Dunbar Cave would lay dormant for several more years before real exploration and surveying resumed.

---

6   Steve Reesman, "Clarksville, Tennessee Trip Report," *The Michiana Caver*, vol 12, no. 11, pp 73, 80–81.

7   Anonymous, "Dunbar System," *The Michiana Caver*, vol 13, no. 2, p 18.

---

8   Steve Reesman, "Woodard Gating Project," *The Michiana Caver*, vol 13, no. 5, pp 47–48.

Chapter 23

Double Dead Dog Drop

Considering the luck they had locating the Roy Woodard Entrance to Dunbar Cave, the Northern Indiana Grotto routinely looked for additional entrances into the Dunbar Cave System on the sinkhole plain that lay to the north of the known extent of the cave. Mark Deebel describes such a trip that occurred on January 22, 1993:

> As some of you know from recent articles in the newsletter, during the traditional Thanksgiving trip to the Dunbar–Woodard Cave System it appeared as though people have gotten into Woodard cave within the last year, and they are obviously not big on conservation. This was my first time in Woodard (Thanksgiving 1992), since I had just recently joined the grotto, and when the January trip was scheduled for Woodard to search for the second entrance no one had to twist my arm to get me to go again. Anyway, on Friday, January 22, I rode down with Bruce DeVore and Floyd Waldrop and we arrived in Clarksville somewhere around midnight. There we met Clyde Simerman, Gene Pelter, Steve Vaughn, and Jack Countryman at the Motel 6 where we roughed it for the night.
>
> Steve, Clyde, and Gene had arrived Friday afternoon and spent the day cleaning out the infamous crawlway, and then complained to anyone who would listen about their sore chests, although they did do a nice job making the crawlway wider (just ask Bruce). On Saturday we all congregated at Dunbar and waited for a few other people to show up. Things started slowly, but we finally decided to drive around and check out the location of a few sinks nearby. After tramping through the woods for a while without finding much, Bruce, Floyd, and I parked along the road for some reason and went walking through a field with some obvious sinks in it. They went one way, so I decided to go the other. I headed towards a large sink while they checked out a few others. As I got closer it appeared as if there was nothing special about it, however, as I reached the bottom I found what appeared to be a pit.
>
> It was a decent-sized opening, and as I leaned over the edge, I couldn't see the bottom. I went back to find Bruce and Floyd, and after showing them the pit they agreed it needed further exploration. Bruce got out the vertical gear and rigged to a nearby tree and Floyd was the first to go down. Steve and Gene happened by about this time and we waited for Floyd to come back up with some news about the pit. The first drop turned out to be about 20 feet deep, followed by an 8-foot drop set off from the first one, which was followed by another 20-foot drop. Bruce went down next and found out that the walls of the pit are very unstable. Gene

went down after Bruce was finished and made it to the bottom of the last drop. (It should also be noted that there is a significant archaeological find at the bottom of the pit—namely, two dog skeletons complete with collars—hence the name of the pit. One even attacked Gene[1] while he was down there—probably just defending its territory.) During this time everyone else had gathered at the site, and we were getting curious stares from people driving by along the road. Even though it was the middle of January, it was a beautiful day, which made the wait at the top of the pit a little more enjoyable. Everyone agreed that the pit needed further exploration, but since it looked like few people, if any, had been down there it probably wasn't the second entrance to Woodard (although rumor has it that it may connect to the Mammoth system). It was about three in the afternoon now and several people wanted to go caving. Those of us who could fit through the crawlway (the Woodard Entrance Crawl) had an enjoyable trip back to a 25-foot waterfall, but saw no signs of a second entrance. All in all it was an enjoyable trip—at least we can say we know where the second entrance isn't located, if nothing else.[2]

Due to their finds in this cave, the explorers named it Double Dead Dog Drop, or D4, for short.

The June–July 1993 issue of *The Michiana Caver* reports a second trip into Double Dead Dog Drop on April 24–25, 1993, by Mark Deebel, Steve Vaughn, Steve McKenzie, and Mike DeVore.[3]

---

1   Dan McDowell, Personal communication, March 28, 2005: Gene Pelter had an accidental fall while exploring this pit, which resulted in a facial cut that required stitches.

2   Mark Deebel, "Double-Dead-Dog-Drop," *The Michiana Caver*, vol 20, no. 2, p 12.

3   Anonymous, "Wayward Wanderings," *The Michiana Caver*, vol 20, no. 4, p 30.

* * * * * * * * * *

It is not uncommon for different groups of cavers to "discover" the same cave over a period of years. As luck would have it, a group of Clarksville cavers re-discovered Double Dead Dog Drop just two years later, in 1995. Charles Blakeway was contacted concerning a newly located cave entrance and he recalls the trip to follow up on that lead:

In 1995, a friend of mine who worked with me at the Clarksville–Montgomery County Rescue Squad, told me of a cave entrance in a sinkhole in a newly developed subdivision that was close to Dunbar Cave. His name is James Duckworth and he knew the developer, who later gave us permission to check out this cave. Since it was a pit, I was very eager to bounce it, since pits are somewhat scarce in this locality.

We set up the trip for a Sunday, although I cannot remember the exact date. The members of the trip were James Phillips (then my shift leader at the Rescue Squad), James Duckworth, and me.

We arrived at the cave and found no trees within 100 yards of the entrance. I drove my Jeep to within 50 feet of the pit, on the uphill side, and rigged to the Jeep. We still use my Jeep for a rigging point to this day, but I park at the bottom of the sink and my Jeep is a newer model.

After gearing up and rigging in, I bounced the entrance, an open-air shaft, which was approximately 4 feet by 6 feet and 30 feet deep. At the bottom, the floor sloped to another drop of about 8 feet and continued into the darkness. Duckworth met me on the bottom of the first drop and I bottomed the next nuisance drop, to discover another 20-foot drop about 15 feet down the passage.

As I shined my light down this third drop, something large, brown, and furry started moving and made whining noises. I yelled back to Duckworth that there was a

Figure 23-1. Matt Simerman in the Entrance to Double Dead Dog Drop.
Photo by James David Reyome, July 14, 2003.

peared emaciated and dehydrated. Duckworth tossed me some extra webbing and I fashioned a makeshift harness in short order.

We hoisted the poor beast up the third and second drops and waited while Duckworth climbed the entrance drop to retrieve some gear. While topside, Duckworth told Phillips about the dog and they threw together a hauling system. They pulled the dog out and then I climbed out to join them. After giving the dog some water, Phillips fed it a sandwich.

Then Duckworth and I re-entered the cave to push the passages at the bottom of the third drop. A crawlway led off from the bottom, but was partially blocked by the remains of another, less fortunate dog. Large passage loomed above an 8-foot-high mud bank that contained lots of layers with lots of bones. This cave has apparently caught more than just dogs over the eons.

On top of this bank was the skeleton of another dog and a leather collar. We saw about 100 feet of passage that pinched out in mud fill. We found no bypass to the grim crawl. Duckworth and I exited the cave and arrived on the surface just as some

dog trapped in the cave. He came down the 8-footer and I rigged in to drop the third pit, with a little apprehension. I rappelled slowly and stopped halfway down. The dog was obviously ecstatic at my presence, but whether it's joy was loneliness or hunger was still a mystery.

Slowly, I finished the descent, keeping my eyes fixed on the dog, a large chow. Arriving at the bottom, I found that the dog was very friendly and in surprisingly good shape for having done three pits by gravity. I found no broken bones, but the dog ap-

kids rode up on bikes. They recognized the dog and said that it had been missing for two weeks. They then pedaled off to call the dog's owner. After 45 minutes they returned and the dog's owner pulled up about 30 minutes after that.

The developer has since built a concrete "bunker" over the entrance to prevent creatures from falling in, especially juvenile primates that inhabit the area. I didn't return to D4 until the connection trip (to Dunbar Cave), but that is another story.[4]

Although we may never know for sure what attracted so many dogs to this pit, in all likelihood they were chasing a raccoon. There have been a number of instances that I know of personally where coon dogs have chased raccoons into caves, become lost, and had to be rescued by cavers. If a coon dog will chase a raccoon into a cave, I'm sure the neighborhood dogs will do the same thing. Just the fact that raccoon tracks are common throughout Dunbar Cave indicates that they are using this and other entrances to the cave. The raccoon was familiar with this cave entrance and avoided falling down the first pit, but the dogs in their rush to catch the raccoon fell in and could not climb out.

---

4   Charles Blakeway, Personal communication, August 12, 2004.

Chapter 24

The Dead Dog Connection

One of the reasons that Charles Blakeway had been so interested in Double Dead Dog Drop was its proximity to Dunbar Cave. When the Northern Indiana Grotto cavers plotted the location of D4 with the topographical overlay map prepared by Jim and Shelley Reyome, it appeared that the new cave was 50 to 200 feet away from an area of rimstone dams in the Roy Woodard section of Dunbar Cave. Since cave surveys are usually prepared using hand-held transits, there is a small degree of error, perhaps one percent, associated with these maps. However, that was clearly close enough to suspect that a connection would be possible. The main concern was that although the rimstone dams and D4 were very close together,

Figure 24-1. Rimstone dams in Woodard Extension where the connection was made.
Shown left to right: Richard Geer and Jack Countryman.
Photo by Tom Cussen, April, 1979.

Figure 24-2. The connection team. From left to right: Matt Simerman, Charles Blakeway, Cory Burkeen, Gary Collins, Tom Willoughby, Melissa Simerman, and Larry Nolan. Photo by Clyde Simerman, July 12, 2003.

they were separated vertically by approximately 100 feet.

The first effort to connect the two caves occurred on May 25, 2003. Gary Collins describes that trip as follows:

Tom Willoughby, Buddy Howerton, and I went in the Woodard Entrance on May 25, 2003, to see if we could find a connection into D4. After studying the overlay map of Dunbar, we knew we had to go either west or northwest from the rimstone dams. Once we were about 100 feet above the dams, Tom found a wide and muddy hands-and-knees crawl off to our right that lasted roughly 50 feet before it opened up into walking (wading) passage.

The cave at this point ran east to west. To our right was a large passage, but we could see only 40 feet away a tough, muddy climb up and to the left was breakdown, which looked to be very unstable. Personally, I liked the looks of the climb up better than the breakdown. So, I pulled out a compass to see which direction we needed to go. It would be an understatement to say I was disappointed when I found out that the breakdown on our left was heading west. Then I noticed to the right of the breakdown there was a small pool of water and it appeared that it could possibly lead us either under or around the breakdown. A few feet past the pool of water I could see about 15 feet and called to Tom and Buddy to come check it out also. We were under the breakdown and there was a hole large enough to climb through, which lead to more rimstone dams. Beneath the break-

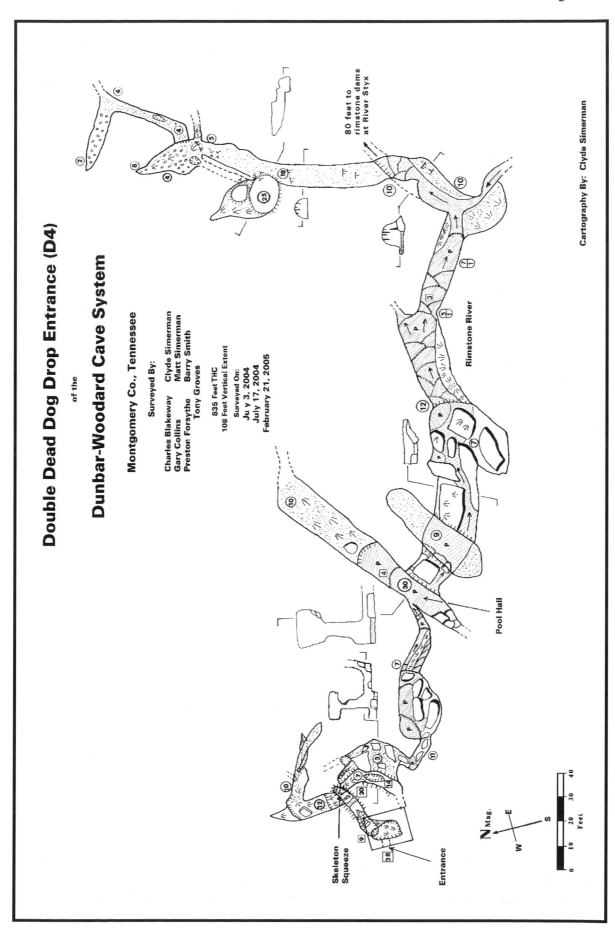

Double Dead Dog Drop Entrance (D4)

of the

Dunbar-Woodard Cave System

Montgomery Co., Tennessee

Surveyed By:

Charles Blakeway   Clyde Simerman
Gary Collins   Matt Simerman
Preston Forsythe   Barry Smith
Tony Groves

835 Feet THC
108 Feet Vertical Extent

Surveyed On:
July 3, 2004
July 17, 2004
February 21, 2005

80 feet to rimstone dams at River Styx

Rimstone River

Pool Hall

Skeleton Squeeze

Entrance

Cartography By: Clyde Simerman

Figure 24-3. Map of the Double Dead Drop Section, 2005.

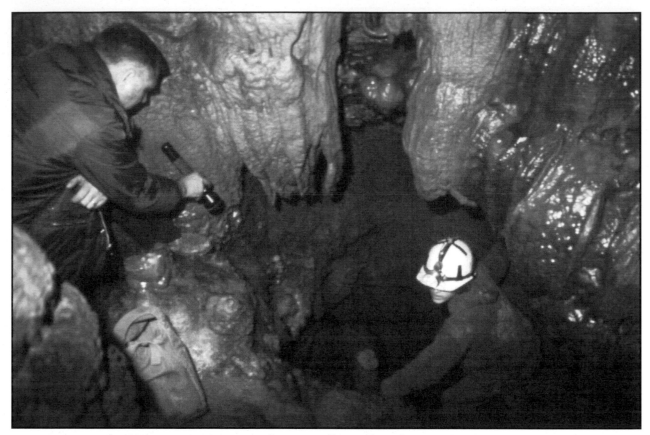

Figure 24-4. Cavers explore a small pit. Photo by Tom Cussen, 1977.

down I stopped to rest, while Tom started sticking his head in every rabbit-sized hole he could find. Then he disappeared in a keyhole above me. He was gone for about 15 minutes. He called back that it was still going and that he found bones. I helped Buddy through the keyhole. The keyhole was at a 45-degree angle and covered in mud. I tried twice, but failed to fit through it. I waited for 30 minutes while Tom and Buddy explored beyond the breakdown.

When they returned, Tom was still carrying a turkey-tail arrowhead in his hand that he had found lying close to bones and leaves. He also mentioned finding small domes. Once I heard there were domes, bones, leaves, and more passage, I felt they had found the connection to D4.

At this time I started sending out e-mail messages about what we had found. I telephoned Gene Pelter and Charles Blakeway, who I knew had been in D4 before, to try to confirm if we had made the connection or not. We were still a little uncertain, but the fact that we had found leaves, which rot fairly easily, and bones that were reported on the D4 side of the cave, made me feel we were very close to the pit entrance.

Because Tom Willoughby had found the keyhole and Charles Blakeway was one of the few people I knew who had been in D4, I wanted to schedule a trip where both of them could go. Since the Northern Indiana Grotto has been instrumental in the exploration, mapping, and surveying of Dunbar Cave for the last 25 years, we wanted to provide enough time for the Northern Indiana Grotto members to be able to join us. The date was finally set for July 12, 2003.[1]

Thanks to the thoughtfulness of Gary Collins, all the major explorers of this portion of

---

1 Gary Collins, "Woodard/Double Dead Dog Drop Connection," *The Michiana Caver*, vol 31, no. 4, pp 24–25.

the cave would be given an opportunity to be on the trip to attempt an actual, physical connection. He continues his story:

Between 10:30 and 11:00 A.M. on July 12, everyone started arriving at the D4 sinkhole. Mac Norman (one of the five owners) stopped by to pick up the release form.

By 11:30 A.M., Cory Burkeen, Clyde Simerman, Matt Simerman, Melissa Simerman, Tom Willoughby, and I headed into the Woodard side of Dunbar Cave. Two hours later we had reached the keyhole above the rimstone dams. Within minutes, we started getting reports that we had made voice contact. I was the last person to go through the keyhole and before I was all of the way out on the other side, Larry Nolan was already waiting on the Woodard side of the dig site. Tom Willoughby found a second turkey-tail arrowhead.

While the six of us were making our

way to the keyhole, Charles Blakeway, Preston Forsythe, Chad Morgan, and Larry Nolan began to work at getting the metal grate opened. Once they had rappelled to the bottom, they found it to be filled with gravel. Charles remembered the spot where he had seen a crawlway five to six years earlier. The four took turns digging for two hours, while they slowly removed about 4 feet of gravel. Some thought there might be better places to dig, but Charles insisted that he remembered exactly where the crawlway was and he was right. The only thing that prevented Charles Blakeway from being the first person to make the connection was the body of a dead dog, which was lying directly in front of the crawlway. He would have had to stick his face over the rotting dog to go through. Charles jokingly commented, "I'm just not hardcore enough to crawl through that." Of the four who helped dig open the crawlway in D4, Charles Blakeway, Chad

Figure 24-5. Caver admires rimstone dams. Photo by Tom Cussen, 1977.

Morgan, and Larry Nolan decided to complete the through trip and come out the Woodard Entrance. Everyone who entered on the Woodard side exited on rope at D4. So, not only were we able to participate in the first Woodard–D4 through trip, but it was the first crossover trip as well.[2]

It had taken eight years, but Double Dead Dog Drop had finally been connected into the Dunbar Cave System. Equally important was the fact that the entrance was securely gated and the cavers had excellent relations with the property owners.

---

2 *Ibid.*

Chapter 25

Modern Cave Tours

After Dunbar Cave was purchased by the state in 1973 to become a part of the Natural Areas Program, it was slowly cleaned and restored and in 1983 the state resumed guided tours of the cave. These tours can best be described as rustic, since the path that is followed is not a concrete walking trail, but is the natural dirt floor of the cave. The persons on the tour will follow the traditional tour that was given during the cave's commercial heyday, but there are currently no electric lights. At one time Dunbar Cave did have an electric light system, but that was removed to restore the cave to a more natural state. Each person is expected to

Figure 25-1. The office for Dunbar Cave State Natural Area was once the Bathhouse for the swimming pool. Photo by Larry E. Matthews, February 15, 2005.

Figure 25-2. A sign points the way to the cave. Photo by Larry E. Matthews, March 22, 2011.

carry his own flashlight, which certainly gives the people on the tour a better feel for what it is like to explore a cave, rather than walking through a typical, brightly-lit commercial cave. There is a modest admission fee for these tours. Tours tend to fill up quickly, so it is always a good idea to call the Park ahead of time and make a reservation.

There are two bridges across the River Styx and steps remain in a few areas. Other than these improvements, the visitor has the feel of exploring a wild cave while on the tour. There are a few low spots, and you will need to be careful to keep from bumping your head on the ceiling. Those ceilings are solid rock and they do hurt, if you hit your head on them. The floor can be muddy at times, so sturdy footwear is recommended and you will not want to wear shoes that will be ruined by a little mud. Also, Dunbar Cave is a steady 56 degrees year round; so many people will want a light jacket

or sweater while on the tour.

Each tour is led by a member of the Park staff. These people are very knowledgeable about the history of the cave, its biology, and the geology. They will share this knowledge with you as they lead you through the cave and they are always willing to answer questions. Dunbar Cave is especially popular with school groups, and thousands of Tennessee school children visit the cave every year. Big, yellow school buses are a regular fixture in the parking lot on most week days.

After the discovery of the Indian glyphs in Dunbar Cave on January 15, 2005, the state decided to make this remarkable cave art available for viewing by the public. Its discovery was announced on July 29, 2006, and all tours, beginning that day, showed the cave art to the public. A barrier, consisting of poles and a chain, mark the limit of how close the visitors may come to this rare and priceless art. Fortunately, the most spectacular and interesting portions of the Indian art are on the main tour.

In 2009, cave tours were conducted Wednesday through Sunday from June until mid-August and on weekends during the spring and fall. Each group was limited to 20 people and reservations were required. The cost was $5 per person. All participants were required to be at least three years old and able to walk for one and a half hours without being carried. Each participant was required to bring a flashlight with fresh batteries.

In March, 2010, a single solitary bat was discovered with White Nose Syndrome (WNS) in Dunbar Cave. Based on this information, the state of Tennessee announced on March 24, 2010, that Dunbar Cave was closed to all visitors and tours were discontinued. However, no other commercial caves in the United States have been closed to tours at this time, even including Howe Caverns in New York where White Nose Syndrome was first identified in February, 2006.

Since bats are social animals and freely leave the cave through its bat-friendly gate, they are free to mingle with bats from other caves and also can, and do, move from cave to cave. Also, most bats migrate and many travel across several states each year. There is little, if any, reason to believe that discontinuing public tours at Dunbar Cave will in any way prevent the spread of White Nose Syndrome.

Hopefully, the state will reconsider its position on public access and reopen Dunbar Cave to the public. However, at the time this book went to the printers, Dunbar Cave was still closed.

Figure 25-3. Amy Wallace stands by the new entrance gate.
Photo by Larry E. Matthews, September 26, 2006.

Dunbar Cave

Chapter 26

The Geology of Dunbar Cave

Since Dunbar Cave is developed in rock, a basic understanding of the geology of the cave and how it formed should greatly improve your appreciation of the cave. Let's consider the type of rock in which the cave is located. Dunbar Cave (and all major caves in Tennessee for that matter) is formed in limestone. Limestone in middle Tennessee tends to be a hard, grayish rock that occurs in horizontal layers, called "beds." Almost every road cut in this area will show you a cross section of limestone beds. Limestone is the rock that is quarried locally and crushed to make gravel, so if you live in middle Tennessee, you see limestone rock outcrops and limestone gravel almost everywhere you go.

Limestone is a sedimentary rock. The term sedimentary rock simply means that these rocks form from material deposited by wind or water. For example, another sedimentary rock known as sandstone forms when wind or water has accumulated beds of sand grains that become buried and cemented together to form a hard rock. When deposits of clay particles become buried and cemented together to form a hard rock, these sedimentary rocks are called shale. Limestone, however, forms in clear, warm tropical seas, such as occur in the Caribbean. If you were to travel to The Bahaman Islands today, for example, you could observe limestone forming on the floor of the ocean. The source material for the limestone is the slow accumulation of seashells. Occasionally, the mineral calcite precipitates directly from ocean water and accumulates on the ocean floor. If you look carefully at the limestone rocks in middle Tennessee, you will frequently see fossil shells, coral, and other traces of animals that lived in those tropical seas.

Now you are probably asking yourself, "How could there have been a *tropical* sea over Tennessee? Tennessee is in the temperate zone, not the tropical zone." The answer to that question is what geologists refer to as "continental drift." The continents are not fixed in place. They "float" on the lower layers of the Earth and can actually move as much as a couple of inches each year. So, over hundreds of millions of years they have moved thousands of miles. When the limestone deposits of middle Tennessee formed during a time known as the Mississippian Period, approximately 320 to 360 million years ago, what is now Tennessee was located very close to the equator. (See Figure 26-1) The particular beds of limestone in which Dunbar Cave are formed are known to geologists as the St. Louis Limestone and the Warsaw Limestone.

What makes limestone so uniquely suitable for the formation of caves is that it is slightly soluble in water. We are all familiar with items that are highly soluble: sugar and salt, for example. It you place a teaspoon of salt in a glass of water and stir for 30 seconds, it disappears. A chemist would say, "The salt went into solution." If you put that same glass of water in a pan on the stove and boiled it until the water all evaporated, the salt would stay behind as a

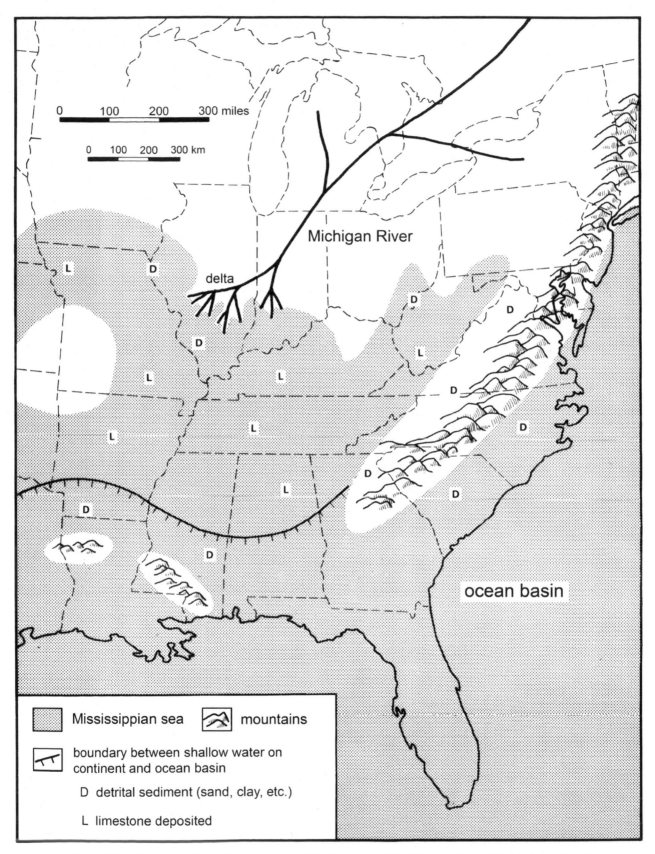

Figure 26-1. This is how the southeastern part of North America looked during the late Mississippian Period, when the limestone rocks of the Dunbar Cave area were deposited. The area that is now Tennessee lay about twelve degrees south of the equator at that time. Map courtesy of Arthur N. Palmer.

deposit on the bottom of the pan. So the salt doesn't "go away," it just goes from a solid form to a dissolved form and back to a solid form. If you took a drink of water from the glass, you would know instantly that the salt was in the water.

If you tried the same experiment with grains of limestone rock, you could stir the glass all day and you wouldn't see any change. It would take months or even years for the same amount of limestone to dissolve, and you would have to use much more water. That is why we say limestone is only *slightly* soluble. However, when you have thousands and thousands of years available, the water naturally moving through the Earth's crust can dissolve and carry off the millions of tons of limestone necessary to create a large cave like Dunbar Cave.

Limestone is barely soluble in pure water,

but *pure* water is not present in nature. Rainwater dissolves carbon dioxide from the air, which makes a mild solution of carbonic acid. As the rainwater filters through the soil, the water becomes more acidic by picking up organic acids and more carbon dioxide. Therefore, by the time rainwater become groundwater, it is mildly acidic and can more easily dissolve limestone and form caves.

Although the area that is now middle Tennessee was underwater for at least 400 million years, it rose above sea level approximately 65 million years ago and has been above sea level ever since. During this time, erosion has worn down the surface to its present shape. But also, slightly acidic groundwater has been dissolving away the limestone to form the many caves that we have now. This water initially moves down the cracks in the limestone (geologists refer to

Figure 26-2. The view from inside the bottom of the entrance, looking out through the arches at the lake. Photo by Larry E. Matthews, March 22, 2011.

177

these as "joints"), then moves out horizontally along the spaces between the beds of limestone (geologists refer to these as "bedding planes"). Naturally, the water follows the easiest path and with time the water slowly dissolves the limestone and forms cave passages. Much of this solution to form large caves goes on beneath or close to the water table.

Once cave passages develop to the point that there is significant stream flow underground, physical erosion can also occur and speed the growth of the cave passages. As the surface streams cut downward, the cave passages eventually become higher then the surface stream level. Then the water drains out of the cave passages leaving behind an air-filled cavity that we call a cave. The water that flows out of the mouth of Dunbar Cave today still carries dissolved limestone. Dunbar Cave is still growing, and will continue to grow, so long as an active stream flows through the cave.

Once a cave is above the local water table, a different set of geologic processes begins. Rainwater moving through the overlying limestone rock still dissolves the limestone. But now, when this limestone-enriched water enters an air-filled cave, it may hang on the ceiling long enough to leave behind a small trace of the mineral calcite. Over a period of many years, this process can form the icicle-shaped formations that we call stalactites. Or the water may drip onto the floor and form cone-shaped mounds that we call stalagmites. If the two formations join, the resulting formation is called a column.

There is a wide variety of formations that form in caves and there are a number of books available on this subject.

All caves have a life cycle. They begin to form beneath, or near, the water table. They become air-filled as the surface streams cut their channels deeper. Then, as the entire land surface is lowered by erosion, they eventually disappear. But this process takes many millions of years. Dunbar Cave has been forming for several million years and it may be here for several million more years before it completely erodes away.

Suggested Reading

Arthur N. Palmer wrote a wonderful book titled *A Geological Guide to Mammoth Cave National Park*. The Zephyrus Press, Teaneck, New Jersey, published this book in 1981. It is currently out-of-print, but it is easy to buy in the used book market and is available in many libraries. Since Mammoth Cave is only 80 miles northeast of Dunbar Cave, the geology of the two caves is very similar and this book is an excellent introduction to how caves are formed in this area.

Another excellent book is *Speleology – The Study of Caves*, written by George W. Moore and G. Nicholas Sullivan. This book was also published by the Zephyrus Press and was first published in 1964. This book has been reprinted and is readily available. It covers both the geology and biology of caves.

Chapter 27

The Biology of Dunbar Cave

The animals that live in caves are generally divided into three groups depending upon their degree of specialization to subterranean life. Animals that are so specialized that they can only live in caves are referred to as *troglobites*. Animals that spend part of their lives underground and part on the surface are referred to as *trogloxenes*. Animals that can live in caves, but also live in cool, moist habitats on the Earth's surface are referred to as *troglophiles*. Animals from all three of these categories live in Dunbar Cave and are discussed below. Other animals occur in caves either by accident, or do not fall in one of the above categories.

Troglobites

The River Styx is home to the southern cavefish, *Typhlichthys subterraneus*. These small, white, eyeless fish average only 1.5 inches long and occur in small numbers. The River Styx is also home to a blind crayfish, *Orconectes pellucidus*. These small, white, eyeless crayfish average only 1.5 inches long and also occur in small numbers.

Dunbar Cave is the type locality for a beetle named *Pseudanophthalmus ciliaris ciliaris*, which was identified by Valentine in 1937. Since Dunbar Cave is the "home" of this beetle, I suggest that we refer to it as the Dunbar Cave Beetle. This is one of a variety of cave beetles found in Tennessee caves.[1] There is also a white millipede, *Pseudotremia sp.*, found in Dunbar Cave.[2] Another small animal, identified as an isopod from the family *Caecidotea* has been identified in the cave.[3] The casual cave visitor rarely sees any of these cave invertebrates, because they are very small.

Trogloxenes

Cave crickets are common throughout the cave. They are most likely *Ceuthophilus gracilipes*.[4] Cave crickets exit the cave through numerous small cracks and crevices that lead to the surface on nights when the weather is mild and similar to that found inside the cave. Once on the surface, they feed on various plants. When they return to the cave, their eggs and feces become an important link in the underground food chain. Some species of cave beetles feed almost exclusively on cave cricket eggs. When a cricket dies, its remains are a virtual feast in this food-poor environment.

Bats are occasionally seen in both the Historic Section and the New Discovery portions of Dunbar Cave. They are generally solitary bats and no bat colonies occupy the cave at the present time. Amy Wallace, Interpretive Specialist at the Dunbar Cave State Natural Area, reports that during the summer of 2004 about

1 Thomas C. Barr Jr, Personal communication, September 30, 2004.

2 *Ibid.*

3 Amy Watkins Wallace, personal communication, December 13, 2004.

4 Thomas C. Barr Jr, Personal communication, September 30, 2004.

*Figure 27-1. Blind cavefish, Hidden River Cave, Kentucky.
Postcard from about 1940. Larry E. Matthews Collection.*

25 to 30 bats came out of the Historic Entrance each night. They are eastern pipistrelle bats, *Pipistrellus subflavus*. She also reports an occasional sighting of the big brown bat, *Eptesicus fuscus* and the northern long-eared bat, *Myotis septentrionalis*. The little brown bat, *Myotis lucifugus*, has been observed near the Woodard Entrance[5]

It is apparent, however, that a large colony of bats did inhabit Dunbar Cave at some time in the past. There is a large area of ceiling in the main passage just before the Stone Mountain Room that exhibits the characteristic reddish-brown stain left when tightly packed bats numbering in the thousands urinate upwards. Over a long period of time, the bat urine reacts with the limestone ceiling to leave this deposit, which is the mineral francolite, a complex calcium fluoride phosphate. Due to the proximity of the Red River and the Cumberland River, the bat colony may have consisted of gray bats, *Myotis grisescens*. Whatever type of bats they were, they probably would have abandoned the cave once it became commercialized in the mid 1800s.

In June, 2006, a "bat friendly" steel gate was built to replace the old gate from the commercial cave era. A bat friendly gate has bars that are horizontal and spaced far enough apart to allow bats to easily fly in and out of the cave. The old gate had bars that were placed vertically, making it more difficult for bats to fly in and out. It was hoped that this improvement would encourage more bats to use Dunbar Cave.

The most frequently observed bats in Dunbar Cave are eastern pipistrelles (80 to 90%), little brown bats (5%), and big brown bats (less than 5%), with a few other species being seen from time to time.

Dunbar Cave, Bats, and White Nose Syndrome

In March, 2010, a single solitary bat was discovered with what were believed to be symptoms of White Nose Syndrome (WNS) in Dunbar Cave. Based on this information, the Tennessee Department of Environment and Conservation announced on March 24, 2010, that Dunbar Cave was closed to all visitors and tours were discontinued.

The bat that was suspected of being infected with WNS was a long eared bat. Perhaps the

5   Ed Vengrouskie, "The biological survey begins at Roy Woodard," *The Michiana Caver*, vol 6, no. 7, pp 57, 64.

Figure 27-2. Cave salamander, Eurycea lucifuga.
Photo by Bob Biddix, July 28, 2007.

fact that the bat was sick had caused it to stop here during migration, or perhaps it was disoriented and unable to follow its normal route. At any rate, the point is that sick bats can, and do, move from cave to cave, and are undoubtedly the vector that is spreading this disease from one cave to another. It is unknown at this time if this single, sick bat transmitted the disease to the small number of bats that live at Dunbar Cave.

The spread of WNS has followed normal bat migration paths and all scientific data strongly suggests that the disease is spreading naturally through the bat population of the United States and that its spread is in no way related to human visitation of caves, especially commercial caves. The disease was first noticed in the state of New York in February, 2006,

Figure 27-3. Cave salamander, Eurycea lucifuga.
Photo by Bob Biddix, August 16, 2004.

and has moved steadily south and west several hundred miles each year. There is no apparent correlation between the presence of the disease and whether or not a cave is open to the public.

Troglophiles

The cave salamander, *Eurycea lucifuga*, is routinely observed in the cave. This beautiful salamander is orange with black spots and is up to 6 inches long.[6] Other salamanders have also been observed in the cave, including the slimy salamander, *Plethodon glutinosus* and the long tail salamander, *Eurycea longicauda*. Like cave crickets, these animals may migrate to the surface to feed when the weather conditions are favorable.

Figure 27-4. Slimy salamander.
Photo by Bob Biddix, August 6, 2005.

The Allegheny wood rat, *Neotoma magister*, is also found in the cave. It occurs so frequently in caves that it is also referred to as the cave rat. However, this is a cute little animal that looks more like a hamster than a rat.

Other Animals

One of the animals that frequents the New Discovery portions of Dunbar Cave is the raccoon, *Procyon lotor*. Explorers have frequently seen raccoon tracks in this portion of the cave. Raccoons have the uncanny ability to wander deep into caves and find their way out. No one is really sure how they do this, but they clearly are not lost, or dead raccoons would be found from time to time. Studies by biologists in

6   Ed Vengrouskie, *op. cit.*

Figure 27-5. Cave rat, Neotoma magister, *in nest. Photo by Bob Biddix, December 3, 2005.*

other caves indicate that raccoons are locating, catching, and eating crayfish and salamanders in the dark. This is a common practice for raccoons through the southern United States and is not unique to Dunbar Cave.

From time to time, ordinary surface crayfish are also seen in the cave's stream. These apparently wash in from the outside or move upstream from the lake and can live for some time in the cave. However, since they are not well adapted to life in a cave, they eventually starve to death and do not reproduce.

It is also possible for regular surface varieties of fish that live in the lake to swim into the cave and be present in the River Styx where they can live for some time. However, like the surface crayfish, they are not well adapted to life in a cave and will eventually starve to death and do not reproduce in the cave.

Phoebes regularly build nests inside the twilight zone of the Dunbar Cave entrance. These fascinating birds are easily observed during the summer months. A fast-moving bird is seen at the upper right edge of the cover photograph of the First Edition (2005).

Some of the more unusual animals observed in the cave include opossum and beaver. They may have wandered into the cave to escape extreme cold weather outside. Several snakes have been observed in the twilight zone, including the rat snake and the ring neck snake. Pickerel frogs have also been observed in the twilight zone. A moth, possibly *Scoliopteryx libatrix*, has been observed hibernating in the twilight zone.

Suggested Reading

Charles E. Mohr and Thomas L. Poulson produced a wonderful book on these animals, titled *The Life of the Cave*. McGraw-Hill Book Company published this book in cooperation with *The World Book Encyclopedia* in 1966. The book is part of a series of books, called "Our Living World of Nature." This book contains exquisite color photographs of cave life accompanied by a text that is easily read by the non-scientist. Although this book is out of print, you may be able to find it in your local library. If not, copies are usually available at online book and auction sites.

Figure 27-6. Brown bat. Photo by Bob Biddix, December 27, 2010.

Chapter 28

Other Natural and Historical Attractions Near Dunbar Cave

Visitors to Dunbar Cave may wish to visit several other interesting natural and historical attractions that are nearby. These are arranged alphabetically below:

Bell Witch Cave

Bell Witch Cave is a privately owned cave that is open to the public. The mouth of the cave is an impressive 60 feet wide and 30 feet high with a small stream flowing out and into the nearby Red River. The main level of the cave is 520 feet long and averages 8 feet high and 6 feet wide. A small upper level is present, but is not usually shown on the tour.

The Bell Witch was a mysterious spirit, or ghost, that plagued John Bell and his daughter Betsy from 1817 to 1828.

Figure 28-1. Carol Lavendar stands by the sign pointing the way to the Historic Bell Witch Cave. Photo by Larry E. Matthews, March 29, 2011.

Figure 28-2. Sign for the Historic Bell Witch Cave.
Photo by Larry E. Matthews, March 29, 2011.

President Andrew Jackson visited the Bell home and came away a believer that the ghost was real. The old Bell Mansion was located about half a mile southeast of the cave. In the particular legend in which the cave is featured, young Betsy Bell and some of her friends had gone to spend the day underground in the cave. One of the boys crawled headfirst into a hole and became stuck fast. Suddenly a voice cried, "I'll get him out!" The cave was flooded with light, the boy felt hands grasping his feet, and he was jerked out of the hole. The loquacious "Spirit" then proceeded to box the ears of the unwary explorer and delivered a lecture on reckless cave exploring.[1]

Many people witnessed the actions of the Bell Witch and even talked to it. It claimed to be the ghost of a neighbor woman named Kate Batts. The ghost was determined to kill John Bell and to prevent John's youngest daughter, Betsy, from marrying a neighbor boy named Joshua Gardner. It succeeded on both counts. John Bell died on December 20, 1820, and in March of 1821 Betsy broke off her engagement with Joshua Gardner.

Daytime tours of the cave are currently available from 10:00 A.M. until 5:00 P.M. The cave is open from May through October, but schedules may vary. It would be advisable to call ahead for the current schedule at: 615-696-3055.

Fort Donelson National Military Park

Fort Donelson National Military Park is actually about 35 miles west of Clarksville, but if you have the time, it is well worth a visit. Allow an hour to drive from Clarksville to the Park,

---

1   Thomas C. Barr Jr, *Caves of Tennessee*, Tennessee Division of Geology, Bulletin 64, 1961, pp 394–396.

Figure 28-3. Carol Lavendar stands by the sign at the entrance to Fort Donelson National Battlefield.
Photo by Larry E. Matthews, May 27, 2011.

Figure 28-4. Carol Lavendar by one of the Civil War cannons at Fort Donelson National Battlefield.
Photo by Larry E. Matthews, May 27, 2011.

Figure 28-5. Carol Lavendar by a large Confederate cannon aimed at the Cumberland River, Fort Donelson National Battlefield. Photo by Larry E. Matthews, May 27, 2011.

Figure 28-6. Historical marker for Jefferson Davis Birthplace. Photo by Larry E. Matthews, March 29, 2011.

at least two (2) hours at the park, then an hour to drive back to Clarksville. So, at a minimum this is a half-day trip. Better yet, just make this a day trip, and take in the quaint little river town of Dover, too.

Fort Donelson was built by the Confederate States of America to control traffic on the Cumberland River. Nashville, the capital of Tennessee, lies upstream from Fort Donelson, so control of the river was vital to the defense of this town. Fort Donelson was the site of a decisive battle fought from February 11 to February 16, 1862. Federal forces, led by Ulysses S. Grant, forced the Confederate forces to surrender. With the Federal forces moving up the Cumberland River, the Confederate forces evacuated Nashville on February 23, thereby losing both the capital of Tennessee and an important industrial center. Tennessee would never again be under Confederate control.

Fort Henry was located 12 miles to the west on the east bank of the Tennessee River. Grant captured Fort Henry on February 6, and then sent his regular army overland to at-tack Fort Donelson, while the Navy gunboats moved down the Tennessee River, then up the Cumberland River to attack Fort Donelson from the river. Fort Henry is completely under water, flooded by the construction of nearby Kentucky Dam.

Jefferson Davis State Historic Site

Jefferson Davis State Historic Site is located approximately 25 miles north of Clarksville in Kentucky. Jefferson Davis was the President of the Confederate States of America during the Civil War (1861–1865)

This site is the birthplace of Jefferson Davis, who was born here on June 3, 1808. The site features a 351-foot monument to Davis, which resembles the Washington Monument in Washington, D.C. It is the world's tallest concrete obelisk. Construction on the monument began in 1917 and was completed in 1924. There is an elevator inside the monument that will take you to an observation room near the top. This room provides an excellent panoramic view of the surrounding countryside.

Figure 28-7. Sign at the entrance to the Port Royal State Historic Area.
Photo by Larry E. Matthews, March 29, 2011.

Figure 28-8. The only historic building still standing in the Port Royal State Historic Area.
Photo by Larry E. Matthews, March 29, 2011.

Land Between the Lakes

Land Between the Lakes is a huge (170,000 acres) National Recreation Area located about 30 miles west of Dunbar Cave. It is managed by the National Forest Service. It is bounded on the north and east by Lake Barkley (Cumberland River) and on the west by Kentucky Lake (Tennessee River). It covers the northwest portion of Stewart County, Tennessee and portions of Lyon and Trigg Counties in Kentucky.

This is a multiple use area that allows hiking, camping, fishing, and hunting (in season).

Port Royal State Historic Park

Port Royal State Historic Park is located 10 miles due east of Dunbar Cave on the banks of the Red River. At one time there was a quaint covered bridge here, but it has fallen into disrepair.

Port Royal was the site of one of the earliest communities and trading posts in middle Tennessee. It was a longhunter camp as early as 1775 and a trading post by the early 1780s. It was located at the head of the navigation on the Red River and was also on a major stage coach line.

The park consists of 26 acres of land located at the junction of the Red River and Sulphur Fork Creek. Boating, fishing, hiking, and picknicing are all popular activites here. Restrooms are available. There is a Pratt Truss design steel bridge built in 1887 that crosses Sulphur Fork Creek and is still in excellent condition. (See Figure 28-10.)

Figure 28-9. The Jefferson Davis Monument.
Photo by Larry E. Matthews, June 5, 2011.

*Figure 28-10. Carol Lavendar stands on an 1887 bridge crossing Sulfur
Creek at Port Royal State Historic Park.
Photo by Larry E. Matthews, March 29, 2011.*

Chapter 29

Epilogue

Epilogue to the First Edition

It is interesting to look back on the very beginnings of the Dunbar Cave Project. In 1977, Dan McDowell optimistically estimated that there might be as much as 8,000 to 9,000 feet of passages when Dunbar Cave was completely mapped. He mentions that the cave "has been quite thoroughly explored," but then goes on to state, optimistically, that: "Of course, one never really knows what might have been overlooked until the job is begun.[1]" Richard P. Geer goes on to add: "Of course we are all dreaming of the possibility that someone has overlooked something and we will find it, likely of not. And what's wrong with a dream?[2]" Ironically, by 1989 the mapped length of Dunbar Cave stood at 42,594 feet. Little did they suspect how much cave they would discover. And more passage is still being explored and mapped in 2004.

* * * * * * * * * *

The saga of Dunbar Cave continues. Today, Dunbar Cave is the centerpiece of a much-loved park in the heart of Clarksville. Instead of electric lights, visitors to the cave now use flashlights on the guided tours for a much more rustic, wild-cave type of adventure. Al-though they still follow the same "trails" that were leveled and dug in the 1800s for the commercial cave tours, these trails are dirt, and not the well-manicured concrete walkways that are the norm in modern commercial caves. Today's guide is a professional Interpretive Specialist working for the State Parks Program and the information given to the tour group will focus on both the history of the cave and it's biological, environmental, and geological significance.

For many Clarksville residents, the natural areas above the cave are even more important, as access to wilderness disappears and at the same time subdivisions continue to sprawl across the Montgomery County countryside. And, although the golf course is now a separate parcel of land, operated by the city of Clarksville, the golf course and State Natural Area combined provide nearly 500 acres of recreational land available to the public.

Just as important to the public as the cave tours are the various programs offered above ground. These include a bat emergence program during the summer months, bird identification hikes, wildflower identification hikes, and tree identification hikes. Fishing is permitted in the 15-acre lake with the proper Tennessee fishing license. Many people visit the park on their own to feed the birds[3], hike the trails, or just to sit and enjoy the peace and quiet. Numerous school groups also visit the park every year for

---

1   Dan McDowell, "Dunbar Cave—a new project," *The Michiana Caver*, vol 4, no. 3, pp 19, 26.

2   Richard P. Geer, "Comments on the Dunbar Project," *The Michiana Caver*, vol 4, no. 4, pp 29–30.

---

3   Feeding the birds, or any other wildlife, is no longer allowed in the Park.

programs both above and below ground.

Dunbar Cave promises to continue to play a place in Tennessee history and as an important place for recreation for many years to come.

The Idaho Springs, which played such an important role in the history of Dunbar Cave and Montgomery County since the first settlers arrived in 1790 and through the end of the Hotel and Resort era in the 1950s, are no more. A visit to the site of these springs in 2004 located the small retaining walls built around each spring, but the hand pumps and gazebos are no longer there. They sit forlornly in a small, abandoned field, approximately 600 feet southwest of the site of the old Dunbar Cave Hotel and close to Dunbar Cave Creek. In all likelihood, once the hotel burned down, the pumps were no longer maintained. The walls around each spring prevented them from flowing freely on their own, so they have slowly silted up from periodic flooding by the nearby creek. And so one of the most prized and frequently visited sites in Montgomery County for nearly 160 years lies in ruins.

* * * * * * * * *

What will future cave explorers discover in Dunbar Cave? The years from 1978, when the Roy Woodard Extension was discovered until 1989, when the first map was finished, were an era on intense exploration and remarkable discoveries. However, from 1989 until 1995, when the D4 Entrance was discovered, there were few discoveries. Exploration picked up again at that time, but with the connection of D4 to Dunbar Cave in 2003, exploration appears to have slowed again. But cave exploration tends to occur in spurts, rarely in a constant, steady progression.

Dunbar Cave extends northward under a very impressive sinkhole plain. This sinkhole plain extends north for 2 miles to Spring Creek. It is even wider, extending west for 2 miles to the Big West Fork of the Red River and over 3 miles to the east. All things considered, there are at least 10 square miles of sinkhole plain that Dunbar Cave might extend

under. The current explored and mapped passages in Dunbar Cave are all contained within approximately one square mile of this sinkhole plain. From a geological standpoint, there is no reason why many more miles of virgin passageways could not be discovered in the future.

The Northern Indiana Grotto cavers explored a cave in 1989 that they named Swallow Pond Horror Hole.[4] It was their hope that this might prove to be yet another entrance into the Dunbar Cave System. However, in 1989 it was plugged with mud and debris after only 50 feet. Now, in 2004, the Dunbar Cave explorers are digging in this lead in hopes of opening up a connection to yet another undiscovered section of Dunbar Cave.[5]

The key to this and all other exploration, will be for the explorers to maintain access to both the public and private entrances to Dunbar Cave. Anyone wishing to explore Dunbar Cave needs to be aware that all entrances to the cave are gated and the explicit permission of the property owners is required to enter the cave. This obviously includes the Historic Entrance, which is owned by the state of Tennessee.

* * * * * * * * *

Epilogue to the Second Edition

As this book goes to press, Dunbar Cave is closed to the public. Or, at least the Historic Section is closed. The only persons allowed in the Historic Section of the cave are biologists studying the bat population for the possible presence of a relatively new disease in bats, known as White Nose Syndrome (WNS).

Dunbar Cave is a wonderful place and it is greatly loved by many people. For this reason, it is terribly frustrating to see the state Department of Environment and Conservation not provide adequate staff needed to properly administer and maintain the area. However, one must remember that Tennessee has a large (and

---

4   Steve Reesman, "Swallow Pond Horror Hole," *The Michiana Caver*, vol 16, no. 8, pp 90-91.
5   Gary Collins, Personal communication, December 8, 2004.

spectacular) state park system whose funding is totally dependent on the state budget. Many other states fund their park system by charging modest admission fees. However, most Tennessee residents oppose such a system. Perhaps, if they visited some of the parks in other states they would see the wisdom of such a system. It is hard to understand why someone would oppose paying a modest fee to use a state park, but not flinch at the current cost of a ticket to a movie and the price of a soft drink and popcorn. Many Tennessee residents have the strange misconception that parks are free and there is no cost involved in managing and maintain them. Nothing could be further from the truth.

Tennessee actually tried such a fee system a few years ago, where a person could buy a day pass or a season pass to a state park. So many people were opposed to this system, that is was abolished. With so many state parks to choose from, Tennessee needs a season pass that would cover all state parks, for a full year and for a modest fee. Of course, there should be discounts for senior citizens. Without a better source of funding, Tennessee state parks will continue to suffer.

Dunbar Cave was visited by thousands of school children each year. Big, yellow school buses were a common sight in the parking lot. Thousands of other people took the tour every year. No other commercial cave in the state of Tennessee is closed to tourists. Mammoth Cave, just 75 miles away, is also still open to tourists. It is difficult to understand how closing one single commercial cave can have any significant impact on the spread of White Nose Syndrome. Dunbar Cave should be reopened for the enjoyment of the public.

Dunbar Cave

Appendix A

Do You Want To Be A Caver?

Just the fact that you are reading this book and are interested in Dunbar Cave is an indication that you may want to become a cave explorer. Exploring caves can be a lot of fun; but, unfortunately, it can also be dangerous if you are not properly trained and equipped. There is a national organization, the National Speleological Society (NSS), that can provide you with information on local chapters in your area and can put you in contact with local cave explorers. They can be easily reached by e-mail on the World Wide Web at:

nss@caves.org

If you are not a computer user, you can write them at the following address:

The National Speleological Society
2813 Cave Avenue
Huntsville AL 35810-4431

The NSS can also be contacted by telephone at:

256-852-1300

The NSS has a bookstore that sells books on caving and caving techniques. Several suppliers of caving equipment advertise on their Web site and in their publications. Their Web site is located at:

http://www.caves.org/

If you do decide to explore caves, please remember the NSS' motto:

TAKE NOTHING BUT PICTURES
LEAVE NOTHING BUT FOOTPRINTS
KILL NOTHING BUT TIME

Middle Tennessee is located in one of the most cave rich areas of the world. The NSS has active chapters (Grottos) in Chattanooga, Clarksville, Cookeville, Livingston, Murfreesboro, Nashville, Sewanee, and Spencer that you may wish to join, if you live in this area. The NSS Web site can give you up-to-date information on how to contact a Grotto in your area. Never go in a cave without proper training and proper equipment. Some commercial caves offer wild cave tours. These are an outstanding way to learn safe caving techniques from experts before you venture out into wild caves on your own.

Have fun and be safe.

Dunbar Cave

Appendix B

Guide To Place Names

Abyss – The pit blocking the end of the Main Passage. Mentioned in a June 11, 1881, newspaper article.

Acheron – The upstream end of the River Styx, where it emerges from beneath a ledge. This is the headwaters of the River Styx in Greek Mythology.

Acheron Room – A large breakdown room approximately 200 feet downstream from the Acheron.

Aeroplane – Location unknown.

Art Gallery – Location unknown.

Ball Room – Mentioned in Goodspeed's *History of Tennessee* (1886). This is the area on the main passage between the Counterfeiters Room and Concession Rock.

Banana Split (Formation) – This formation resembles a banana split. Location unknown.

Bat Bone Passage – Named for the numerous bat skeletons on the floor, this 1,000-foot-long passage was one of the two major virgin passages discovered in the Historic Section by the Northern Indiana Grotto.

Bear Wallows – Located behind Petersons Leap.

Belfry – Location unknown. Did it have bats?

Bell Witch Chamber – This well-decorated room is in the Hidden Lake Passage, approximately 300 feet northeast of Independence Hall. Probably named by an early Guide in an attempt to scare the tourists. The Bell Witch was a famous "spirit" in nearby Robertson County, supposedly witnessed by President Andrew Jackson.

Blackberry Patch – Location unknown.

Boat – Location unknown.

Bridal Veil – A drapery formation, probably in Independence Hall. Shown on an old postcard. (See Figure 8-5)

Broadway – Location unknown.

Broken Domes – These are several, interconnected domes that have "broken" into each other, due to collapse of the walls. They are near the end of the passage that leads northeast from the Domino Room.

Broken Domes Passage – The passage that leads northeast from the Domino Room to the Broken Domes.

Dunbar Cave

Cactus Formation – A small stalagmite formation resembling a saguaro cactus. Located in Watts Trail.

Cave Cathedral – Is this the same as Cathedral Hall?

Cathedral Dome – This dome is located at the end of a side passage off Watts Trail.

Cathedral Hall – This room consists of several interconnected domes that are 25 to 30 feet high. It is located off Indian Trails Passage near the Jenny Lind signature.

Chief's Wigwam – Location unknown.

Choir Booth – Location unknown.

Concession Rock – A large, flat-topped piece of breakdown along the main tourist trail. Was it once used to sell concessions?

Counterfeiters Room – Named for its alleged use by counterfeiters, this room is located west of the Main Passage. It is 45 feet long, 30 feet wide, and 15 feet high. Shown on an old postcard. (See Figure 6-1)

Crocodile – A formation in Independence Hall that resembles the head of a crocodile with its jaws open.

Crystal Palace – An extremely well decorated room located south of Independence Hall. It is 36 feet high and shows three distinct levels.

Cypress Knees – Location unknown.

D4 – The abbreviated name for Double Dead Dog Drop.

D4 Entrance – The Double Dead Dog Drop Entrance was the third entrance discovered for the Dunbar Cave System.

Dental Factory – Location unknown.

Devils Kitchen – This is a small, flowstone-covered pit in the floor of the passage southeast of Independence Hall. The flowstone has a reddish color. The origin of the name is unknown.

Diamond Grotto – A well-decorated chamber located at the end of the passage that trends southeast from Independence Hall. The name probably derives from the abundance of macrocrystalline flowstone and white soda straws. Shown on an old postcard. (See Figure 8-4)

Discovery Dome – Located at the end of Watts Trail. A scaling pole was used here to enter and explore Upper Watts Trail.

Dog's Head – Located in Independence Hall close to the Monkey and the Pineapple.

Domes Passage – Another name for Indian Trail. Used by the Northern Indiana Grotto cavers in 1977.

Domino Room – This large breakdown room is named for the large slabs of breakdown that are stacked on each other like "dominos."

Domino Room Passage – This passage leads from Grand Avenue to the Domino Room.

Double Dead Dog Drop – This cave entrance, located in 1993, had two vertical drops with several dead dogs at the bottom. It was later connected to the Dunbar Cave System.

Dunbar Cave – Named after Thomas Dunbar, this is the name given to the original cave entrance, discovered in 1790.

Dunbar Cave Entrance – The original entrance to the Dunbar Cave System. It is 12 feet high and 60 feet wide, located at the base of a 75-foot-high bluff.

Dunbar Cemetery – Location unknown.

Dunbar Hill – This very steep, slippery slope in the Indian Trails Passage leads up into the passage that contains Triangle Rock.

Dunbars Coffin – Mentioned in Goodspeed's *History of Tennessee* (1886). Location unknown. Could this be an early name for Concession Rock?

Egg Shell – Location unknown.

Egyptian Grotto – One of the well-decorated rooms on the old commercial tour, located just south of Independence Hall. Probably named for the flowstone columns, which could be considered to resemble Egyptian ruins. Shown on an old postcard. (See Figure 6-2)

Elephant – Mentioned in Goodspeed's *History of Tennessee* (1886). Located in Independence Hall. (See Figure 5-5)

Entrance Room – The first room of Dunbar Cave located about 200 feet from the entrance. It is 150 feet long, 65 feet wide, and 20 feet high in the center.

Everglades of Florida – Location unknown.

Fat Mans Misery – A tight spot in the passage between the Egyptian Grotto and the Crystal Palace. At some point, dirt was removed to make this 10-foot-long slot more easily negotiated by the tourists.

Fluffy Ruffles – Location unknown.

Fossil Ravine – Location unknown. Described as containing "fossils" of alligators, water dogs, eels, turtles, and monkeys.

Fountain of Youth – A small, water-filled alcove on the right wall where you enter the Stone Mountain Room.

Frank Meador Passage – Named for Frank Meador, the owner of the Woodard Entrance. This passage extends east off of Watts Trail, approximately 500 feet past Signature Rock.

Frozen Waterfall – A large flowstone formation located at the junction of the Rimstone River Passage and the River Styx. Also referred to as the Flowstone Waterfall by some explorers.

Grand Avenue – This major passage extends from the River Styx Passage south and then east to the Domino Room.

Great Relief Hall – This was the first place that the tourist could stand up after a lengthy stoopway. It is 125 feet long, 75 feet wide, and 15 feet high.

Gumbo Avenue – This is an alternate name that was used for Grand Avenue.

Happy Dutch Family – Mentioned in Goodspeed's *History of Tennessee* (1886). Located in Independence Hall.

Head of Alligator – Location unknown.

Henry Ford's Factory – Location unknown. Supposed to look like caps for radiators.

Herculanaeum – This name is written in white paint on the wall of the Junction Room.

Hidden Lake – A small pool at the end of a crawlway off the Hidden Lake Passage. This pool is 10 feet deep with a large flowstone canopy hanging over it.

Hidden Paradise Room – A hard-to-find room located just south of the Paradise Room. This is arguably the most beautiful room in the cave.

Dunbar Cave

Historic Section – The portion of Dunbar Cave that was explored from the Historic Entrance.

Hornets Nest – Location unknown.

Horse Skull Passage – This is a side passage leading west from the Woodard Entrance Passage at the point where a horse skull was discovered.

Iceberg – Located in Independence Hall. Probably named after the Titanic sank in 1912.

Ice Pond – This is the name for a portion of the cave floor that is covered with white, macrocrystalline flowstone, which resembles ice. It is located south of Independence Hall. Much of this area has been walked on over the years and is now covered with a thin layer of mud.

Idaho Springs – Three natural sulphur springs that were located 2,750 feet south of Dunbar Cave.

Impression of Giant's Hand – Location unknown.

Independence Hall – This was the main attraction of the commercial tour. It is a large, very well decorated room. The name dates from the 1800s, but its origin is unknown. Beneath the carefully lettered sign, "Independence Hall," written on the roof are the letters: **"O.L.S."** Their meaning is unknown. This room is 80 feet in diameter and up to 18 feet high.

Indian Grave Site – This small alcove is located in the west side of the Main Passage, about 100 feet from the Dunbar Cave Entrance.

Indian Trail – This passage extends northwest from the Main Passage. This is an early commercial name and its origin is unknown.

Irish Potatoes – Mentioned in Goodspeed's *History of Tennessee* (1886). Located in Independence Hall.

J Passage – One of the passages leading off of Independence Hall. It is 400 feet long. Named for its survey designation.

Job's Coffin – Probably the same feature as Dunbars Coffin.

Jacobs Well – Mentioned in Goodspeed's *History of Tennessee* (1886). Located in Independence Hall.

Junction Room (Historic Section) – This room is located in the middle of the Historic Section, with passages radiating out in various directions.

Junction Room (Woodard Section) – This room is located in the Roy Woodard Section of the cave. It is located at the intersection of the Woodard Entrance Passage and the River Styx.

Keyhole – A small hole that connects the end of the Rimstone Dam Passage in the Woodard Extension of the cave to the end of Double Dead Dog Drop.

Lake Passage – This is a room located immediately to the northwest of Independence Hall. It is named for a small pool located in the center. This room is 50 feet in diameter.

Land of Romance – Location unknown.

Lovers Lane – Is this the passage leading to Lover's Leap?

Lovers Leap – In all likelihood, a name dreamed up by an early guide to entertain the tourists. This small drop-off is located just north of the Junction Room. At some point in time, steps were dug in the dirt, so it is now easy to negotiate.

Lower River Passage – This is the portion of the River Styx from the Junction Room in the Woodard Extension downstream to the sump where it connects to the Historic Section.

Madonna and Child – A stalagmite on the northwest bank of the River Styx in the Entrance Room. Shown to tourists on the commercial tour.

Main Passage – The passage leading from the Dunbar Cave Entrance to the Stone Mountain Room.

Mammoth Elephant – A formation that resembles a Wooly Mammoth. Shown on an old postcard. (See Figure 5-5)

Marble Yard – Location unknown. An area with cave pearls?

Maze Area – This area near the end of the Northwest Passage is named due to its maze-like array of passages.

Merry-Go-Round – This consists of several loop passages near the end of Upper Watts Trail.

Milky Way – Location unknown.

Miniature Screech Owl – Location unknown.

Mirror Pool – Probably in Independence Hall.

Monkey – Mentioned in Goodspeed's *History of Tennessee* (1886). Located in Independence Hall. This formation is supposed to resemble a monkey.

Moonshine Room – This was another name for the Counterfeiters Room used at one time during the cave's commercial history.

Music Hall – Mentioned in Goodspeed's *History of Tennessee* (1886). This is the section of the Main Passage next to the Counterfeiters Room.

Mushroom Farm – The Dunbar Cave State Natural Area files contain a photo of a mushroom farm, but the date that this farm was in operation and the location inside the cave are unknown.

Negro and His Coon – Location unknown.

NIMIG Siphon – Another name for the River Styx Siphon. NIMIG stands for Northern Indiana-Michigan Interlakes Grottos.

Northwest Passage – This major passage extends northwest from Dunbar Hill, the end of the old commercial route.

Paradise Room – This large, attractive formation room is located at the end of the Paradise Room Passage and contains many beautiful and colorful formations.

Paradise Room Passage – This passage leads south from Grand Avenue to the Paradise Room.

Perfect Tunnel – Location unknown.

Petersons Leap – The 35-foot drop located immediately northeast of the Junction Room where Bryant F. Peterson fell on July 7, 1844. During commercial times, a mannequin lay at the bottom of this drop.

Petrified Coon – Mentioned in Goodspeed's *History of Tennessee* (1886). Located in Independence Hall. This formation supposedly resembles a raccoon.

Pineapple – Located in Independence Hall between the Dogs Head and the Monkey.

Dunbar Cave

Pipe Organ – Location unknown. Was there ever a commercial cave without a formation named the Pipe Organ?

Pulpit – Location unknown.

Rebeccas Seat – Mentioned in Goodspeed's *History of Tennessee* (1886). Located in Independence Hall. Perhaps someone named Rebecca used it as a chair?

Rebeccas Well – Location unknown.

Relief Hall – Mentioned in Goodspeed's *History of Tennessee* (1886). This may be an early name for the Junction Room, used before the passage leading from the Main Passage to this room had been dug out to walking height.

Rimstone River – The name of the stream that flows through the Rimstone River Passage.

Rimstone River Passage – This passage in the Woodard Extension is named for its numerous rimstone dams.

River Styx – The name given to the main stream that flows through Dunbar Cave. The original River Styx is named in Greek mythology as the river in the underworld over which the dead were ferried on their way to Hades.

River Styx Siphon – The point, approximately 1,000 feet upstream from the Dunbar Cave Entrance, where the cave's ceiling drops lower than the surface of the stream. This is not a true siphon and would more correctly be described as a sump.

Rock of Ages – Location unknown.

Rocky Mountain – Mentioned in Goodspeed's *History of Tennessee* (1886). This is probably the huge breakdown pile at the end of the Stone Mountain Room.

Rocky Mountain Route – This passage is located east of the Main Passage and has considerable breakdown, hence the name. It connects Petersons Leap to the far end of the Stone Mountain Room.

Rodent Dome – This large dome is east of the Horse Skull Passage. It was named for a number of rodent skulls on the floor.

Roy Woodard Cave – This cave, which has been connected into the Dunbar Cave System, was named for a former owner of the property.

Roy Woodard Entrance – The Roy Woodard Entrance was the second entrance discovered for the Dunbar Cave System.

Roy Woodard Extension – This is the portion of the Dunbar Cave System that lies past the River Styx Siphon and is most easily explored by entering through the Roy Woodard Entrance.

Roy Woodard Sink – The deep sinkhole in which the Roy Woodard Entrance is located.

Saint Georges Avenue – A very small side passage southeast of Independence Hall that is 2 feet high, 1 foot wide, and 25 feet long. The origin of the name is unknown.

Saltpeter Mine – Mentioned in Goodspeed's *History of Tennessee* (1886). It was located in the area of the Ball Room along the Main Passage.

Sharks Jaws – This formation resembles the open mouth of a shark. It is located in the Hidden Paradise Room.

Signature Rock – The location in the Roy Woodard portion of the cave where four explorers left their names and the date on October 14, 1939. Located in Watts Trail 250 feet north of Rodent Dome.

Solomons Pool – This feature is shown on an old commercial postcard. Located in Independence Hall. (See Figure 5-4)

Solomons Porch – Mentioned in Goodspeed's *History of Tennessee* (1886). Located in Independence Hall.

Spray Hall – This room consists of several interconnecting domes and is named for its small waterfall. It is located off Indian Trail passage in the Historic Section.

Squaw's Wigwam – Location unknown. Supposed to have the likeness of an Indian Warrior standing in the doorway.

Stone Mountain Room – This is the largest room in the Historic Section. It is named for the large pile of breakdown that occupies most of the room. It measures 200 feet long, 100 feet wide, and 35 feet high.

Surprise Dome – This dome is located on the east side of the Broken Domes Passage, a few hundred feet before the Broken Domes.

Three Story Red House – Location unknown.

Throne Room – This room is located across the River Styx from the Entrance Room in the Historic Section of the cave. During the commercial tour era, there was a bridge across the stream that allowed tourists to visit this room without getting wet.

Tobacco Barn – Location unknown.

Triangle Rock – This is a large, triangular piece of breakdown, located near the end of the Indian Trails Passage in the Historic Section. This was the end of the commercial tour.

Truck Farm with Radishes, Turnips, and Parsnips – Location unknown.

Twin Domes – These domes are located on the west side of Watts Trail, approximately 600 feet northwest of Signature Rock.

Twin Waterfall Domes – This is another name that was used by some of the explorers and surveyors for Discovery Dome.

Up & Down Series – Is this another name for the Bat Bone Passage in the Historic Section?

Upper River Passage – This is the portion of the River Styx from the Junction Room in the Woodard Extension upstream to its end.

Upper Watts Trail – This is an upper level passage that leads off the top of Discovery Dome.

Volcano – Located in Independence Hall.

Watts Trail – Named for 1939 explorer Billy Watts, this is the main passage leading generally northwest from the Woodard Entrance Passage.

Wedding Chamber – This is another name for the Counterfeiters Room. Apparently a couple was married there at some time during the cave's commercial history.

Wheelbarrow Track – Mentioned in an 1869 newspaper article.

White Place – This is an early name for the Paradise Room. The discoverers originally called it the White Place due to the abundance of white formations in this room.

Wigwams Number 1, 2 and 3 – Location unknown.

Willapus Wallapus – Mentioned in Goodspeed's *History of Tennessee* (1886). Located in Independence Hall. This formation was supposed to resemble the Willapus Wallapus, a mythical, ferocious beast.

Dunbar Cave

Woodard Crawl – The very tight crawlway that must be negotiated shortly inside the Roy Woodard Entrance.

Woodard Entrance Passage – This is the passage that leads from the bottom of the Roy Woodard Entrance to the Junction Room, where it connects to the River Styx.

Appendix C

Glossary

Acetylene – A flammable gas (C_2H_2) formed by combining carbide and water.

Aragonite – A mineral composed of calcium carbonate ($CaCO_3$), the same as calcite, but forming orthorhombic crystals instead of hexagonal crystals as calcite does.

Bacon Rind – A thin, banded flowstone formation that resembles a strip of bacon.

Bat – A nocturnal flying mammal of many different species. They frequently spend the daylight hours inside caves and some hibernate inside caves during the winter.

Bedding Plane – The contact between two layers of sedimentary rocks.

Belly Crawl – A crawlway so low that it can only be negotiated flat on your stomach.

Biology – The science that studies the living organisms of the Earth.

Breakdown – A pile of rocks in a cave room or passage, formed by the collapse of the cave roof.

Cable Ladder – A flexible ladder consisting of two small steel cables with small, aluminum rungs. These are available commercially in 10-meter sections. They can be rolled into a small bundle for easy transport through a cave.

Calcite – A mineral composed of calcium carbonate ($CaCO_3$), which is the main component of limestone. Calcite is also the main component of most cave formations, such as stalactites and stalagmites.

Carbide – A man-made chemical, calcium carbide (CaC_2). Reacts with water to form acetylene gas. $CaC_2 + H_2O = C_2H_2 + CaO$.

Carbide Lamp – A lamp, usually manufactured from brass, which combines carbide and water to produce acetylene gas, which burns with a very bright flame.

Cave – A naturally occurring cavity in the Earth's crust, which is large enough to permit human exploration.

Cave System – A group of caves connected by exploration. Frequently used to describe a large cave with several entrances.

Caver – A person who explores caves. This is the term cave explorers use to describe themselves.

Chalybeate – An adjective describing spring water that is strongly flavored with iron salts.

Chert – A microscopically grained quartz mineral (SiO_2) of various colors that forms beds, lenses, and nodules in limestone. Frequently used by prehistoric peoples to make stone tools.

Commercial Cave – A cave that has all or part of its passages developed with walkways, steps, and lights and is open to the public for guided tours. A show cave.

Connection – A naturally occurring passage which, when discovered, connects two caves which were previously considered separate caves.

Crawlway – A cave passage that is so low that cave explorers must crawl on their hands and knees or stomachs.

Crossover Trip – A trip where two parties of cavers enter two different entrances to the same cave system, pass each other in the cave, and exit out the entrance that the other party entered through.

Dome – A vertical passage extending upward from the ceiling of a cave passage or room.

Dripstone – Cave formations composed of calcite or aragonite, formed when dripping water deposits dissolved limestone.

Drop – A pit or a cliff that must be rigged with rope or a ladder to descend.

Entrance – A point where a cave opens to the surface and is large enough to permit human passage.

Fill – Secondary deposits of sediments in a cave passage, usually consisting of clay, silt, sand, gravel, or a combination of two or more of these.

Flowstone – Cave formations composed of calcite or aragonite, formed when flowing water deposits dissolved limestone.

Fossil – The remains or traces of animals and plants that lived in prehistoric times.

Francolite – A mineral composed of a complex calcium fluoride phosphate, generally formed by the reaction between bat urine and limestone.

Geology – The science that studies the structure, history, and origin of the Earth.

Grotto – A term used to describe a single cave room. Also the term used for a chapter of the National Speleological Society.

Gypsum – A mineral composed of hydrous calcium sulphate ($CaSO_4$-$2H_2O$), which forms colorless to white crusts and crystals in a variety of forms.

Gypsum Flower – A flower-like group of curved gypsum crystals.

Joint – A fracture in sedimentary rocks, usually perpendicular to the bedding planes.

Limestone – A sedimentary rock composed primarily of calcite.

Lower Level – A series of cave passages developed at a lower elevation than the rest of the cave.

Macrocrystalline Flowstone – Flowstone that is covered with crystal faces that are large enough to be easily seen by the unaided eye.

Map – A two-dimensional representation of a cave, showing its horizontal extent that may include cross-sectional representation to also show the vertical extent.

Midden – A refuse heap left by prehistoric peoples.

MIG – The Michigan Interlakes Grotto of the National Speleological Society.

National Speleological Society – An organization of cave explorers and speleologists, founded in 1941, to promote the exploration, preservation, and scientific study of caves.

Native Americans – This is the "politically correct" term for the people who lived in North America before the arrival of Europeans in 1492. These people prefer to be known as American Indians.

NIG – The Northern Indiana Grotto of the National Speleological Society.

NSS – The National Speleological Society.

Passage – A horizontal section of a cave.

Pit – A vertical passage extending downward from the floor of a cave passage or room.

Popcorn – An irregular flowstone formation with many rounded, knob-like surfaces.

Rimstone – A raised flowstone edge surrounding a pool of water.

Rimstone Dam – A naturally formed flowstone dam occurring at the downstream edge of a cave pool.

Rimstone Pool – A pool of water surrounded by a raised edge of flowstone.

Rock – The solid material that forms the Earth's crust.

Rock Shelter – A rock overhang that resembles a cave entrance but does not open into an actual cave.

Room – A room-shaped opening in a cave.

Saltpeter – A naturally occurring mineral, calcium nitrate, $Ca(NO_3)_2$, used in the manufacture of gunpowder.

Saltpeter Vat – A vat constructed to leach saltpeter from cave dirt.

Saltpetre – A spelling of saltpeter that was widely used in the 1800s.

Scaling Pole – A strong pole, usually transported in sections, placed at the bottom of a difficult climb to support a rope or a cable ladder at the top of the climb.

Sedimentary Rock – A rock formed by the deposit of water-borne and wind-blown particles or by the accumulation of chemically or organically precipitated materials.

Sinkhole – A surface depression in an area of limestone rock formed either by solution as water drains underground or by the collapse of a cave passage or room.

Sinkhole Plain – An area of relatively flat topography with numerous sinkholes.

Siphon – A section of water-filled cave passage connecting two sections of cave passage with air is frequently referred to as a siphon. However, true siphons rarely, if ever, occur in caves. The more proper name for these water-filled sections of cave passage is a *sump*.

Soda Straw – A thin, hollow stalactite, similar in size and shape to a drinking straw.

Soluble – Capable of being dissolved, usually referring to water.

Speleology – The scientific study of caves, their origins, and their features.

Spring – A place where water naturally flows out from underground onto the Earth's surface.

Squeezeway – A passage so small and tight that the explorers must squeeze to get through it.

Dunbar Cave

Stalactite – An icicle-shaped cave formation that hangs down from the ceiling.

Stalagmite – A conical-shaped cave formation that grows upward from the floor.

Stoopway – A cave passage that is too low to stand up completely, but not as low as a crawlway.

Sump – A section of cave passage that is completely filled with water.

Survey – n: A map prepared by using an accurate compass, a measuring tape, a clinometer, and sketches. v: To prepare an accurate survey or map.

Survey Station – A point in the cave from which compass direction, inclination, and distance are measured to the next point.

Through Trip – A trip where cavers enter a cave through one entrance and exit through another entrance.

Upper Level – A series of cave passages developed at a higher level than the rest of the cave.

Vandal – A person who vandalizes a cave.

Vandalism – Altering a cave, either accidentally or intentionally, by breaking formations, leaving trash, writing on walls, or otherwise changing the original appearance of the cave.

Virgin Passage – Cave passage that has not been entered previously.

Walking Passage – A cave passage large enough to walk without stooping.

Water Table – The horizontal plane beneath which all open spaces in the Earth are filled with water.

WIG – The Western Indiana Grotto of the National Speleological Society.

Appendix D

Chronology

10,000 B.C. – American Indians first visit Dunbar Cave.

1500 – American Indians leave Dunbar Cave.

1741 (January 1) – Thomas Dunbar is born in Lancaster County, Pennsylvania.

1753 – Ann Keys in born in Pennsylvania. She marries Thomas Dunbar.

1755 – Isaac Peterson is born.

1783 (January 20) – Rowland Peterson (son of Isaac Peterson) is born.

1784 (August) – Thomas Dunbar moves to what is now Montgomery County, Tennessee. He does not own any land and is living as a squatter.

1785 (October 11) – Ann Dunbar is born in Montgomery County, Tennessee. She is the daughter of Thomas Dunbar.

1787 – Isaac Peterson receives a 480-acre land grant in Tennessee from the state of North Carolina.

1790 (November 19) – Thomas Dunbar sells eight acres of land next to Dunbar Cave to Isaac Rowland Peterson for twenty pounds (currency).

1791 – Peterson returns and finds Thomas Dunbar living at Idaho Springs.

1792 – Peterson receives clear title to the land and becomes the first owner of Dunbar Cave. Another version says that Robert Nelson became the first legal owner of Dunbar Cave in 1792.

1797 (September 29) – Isaac Peterson is commissioned as a Captain in the Montgomery County Militia Regiment.

Dunbar Cave

1809 – Robert Nelson sells Dunbar Cave to John Newell.

1811 – John Newell sells Dunbar Cave to William Tate.

1814 – William Tate sells Dunbar Cave to Charles Cherry.

1823 – Ann Dunbar (daughter of Thomas Dunbar) dies at the age of 38.

1824 (October) – Thomas Dunbar dies in Montgomery County, Tennessee, at the age of 83.

1832 – F. Whitehill (or Whitfield ?) smokes his name on the ceiling near the Confederate Soldier.

1833 (July 22) – Isaac Peterson dies at age 78.

1835 (November 12) – Rowland Peterson (son of Isaac Peterson) dies at age 52.

1835 – Isaac Peterson, the grandson, becomes the new owner of Dunbar Cave.

1839 – Thomas Cherry (Charles Cherry's oldest son) is convicted of holding and passing counterfeit coins.

1842 – Counterfeiters use Dunbar Cave to store counterfeit coins.

1843 – John Nicholas Barker becomes the new owner of Dunbar Cave. He owns the cave from 1843 to 1868.

1843 – B.E. Ringonelert writes his name on the wall at the Bear Wallows.

1844 – J. Cording and A.L. Bobo smoke their names on the ceiling of Great Relief Hall.

1844 (July 7) – A large party and bran dance are held at the cave. Bryant Peterson falls into the ravine now known as Peterson's Leap.

1846 – W.J. Smoot and A. Sharon smoke their names on the ceiling in Great Relief Hall.

1847 – Dunbar Cave is mined for saltpeter.

1848 – W.H. Sydmor smokes his name on the ceiling of the Junction Room.

1851 – World famous singer Jenny Lind (the Swedish Nightingale) performs in Nashville and may have visited Dunbar Cave.

1851 – The name Jenny Lind and the date 1851 are written on the wall in the Indian Trails Passage.

1855 – Anna Breen writes her name on the wall of the Main Passage between the First Room and the Counterfeiters Room.

1855 – IPCO, BFW, BIC, and HF write their initials and the date on the wall of the Main Passage between the First Room and the Counterfeiters Room.

1858 – Cabins are built at Idaho Springs.

1858 – A.J. Harrison leaves his name and the date on the wall across from the steps leading up into the Counterfeiters Room.

1858 (May) – A.L.H. writes his initials and the date in red chalk on the wall on the main Passage between the First Room and the Counterfeiters Room.

1860 (January 22) – C.W.U. and J.E. smoke their initials on the ceiling in the Junction Room.

1862 – A portrait of a Confederate Soldier is drawn on the wall and the date 1862 is written above it.

1862 (February 22) – Union troops supported by gunboats enter Clarksville.

1865 – G. Holt Argent smokes his name and the date on the ceiling near the top of Lovers Leap.

1866 (July 4) – Picnic held at Dunbar Cave to celebrate the 4th of July.

1867 – G. L. A. smokes his initials and the date onto the ceiling in Great Relief Hall.

1871 – B.B. Avoeurn, Singe, and C.L. Thomas write their names and the date on the wall in black paint about 150 feet into the cave. They also write the phrase: "Thomas' Chill Tonic."

1872 – Belle Waggoner writes her name and the date on the wall of the Indian Trails Passage.

1873 – John Nicholas Barker dies.

1874 (July 29) – Bryant F. Peterson returns to Dunbar Cave.

1874 (October 3) – Sale of 240 acres of land belonging to the late J.N. Barker, which includes Dunbar Cave.

1876 (August 24 – 25) – Montgomery County Agricultural Association meeting held at Dunbar Cave.

1877 (August 5) – L.M.L. and H.K.A. write their initials and the date on the wall in the Ball Room.

1877 (August 5) – W.E. Beach writes his name in black paint on the wall in the Ball Room.

1878 – J. Frisby writes his name in red paint (or chalk?) on the wall of the Main Passage 100 feet past Concession Rock.

1878 (July 21) – Luc. Gaisser writes her name and the date on the wall at the top of Peterson's Leap.

Dunbar Cave

1879 (March 28) – The young ladies of Hickory Wild Academy visit Dunbar Cave with Professor Tate.

1879 (June 21) – Charles Warfield purchases Dunbar Cave from Chancery Court.

1879 – James A. Tate builds the first Idaho Springs Hotel.

1880 – Commercialization of the cave begins. The newspaper announces that admission to the grounds is 10 cents and that guided tours are 50 cents, with discounts to large parties. Barbeque prepared when desired.

1881 (July 14) – John W. Rice writes his name and the date on the wall near the entrance to the Indian Trails Passage.

1881 (August 26) – Miss G. Sriger writes her name and the date on the wall in the Indian Trails Passage.

1882 – J.M. Rice, C.P. Warfield, and J.P. Gracey purchase Dunbar Cave.

1882 – C.P. Warfield wedged in cave by falling rock for 30 minutes.

1882 (May 30) – The newspaper announces that: "Dunbar's Cave is now open to visitors. Season tickets on liberal terms."

1882 – J.T. Wood writes his name and the date in Great Relief Hall.

1883 (July 4) – A bran dance is held at the cave with Dick Jackson as vocalist and Sam Aeglen as instrumentalist.

1883 (August 24) – Montgomery County Agricultural Association meeting held at Dunbar Cave.

1884 – Colonel A.G. Godlett and H.C. Merritt purchase the hotel and cave.

1885 – A.C. Goodlett (age 15) works at the cave hauling dirt out in a donkey cart, while the inside passages were being dug deeper.

1885 (July 5) – C.P. Warfield was host to the executive committee of the Stock Breeders and Farmers Association. He took them on a tour of his farm and then ended up at the cave.

1886 (January 27) – The Nashville *Daily American* reports on the use of "Cave Saltpeter as a Tobacco Fertilizer".

1886 (July 1) – W.J.A. writes his initials and the date with a candle on the ceiling of the passage that connects Great Relief Hall with Independence Hall.

1886 – A large number of small lots are sold for use as building sites for cabins.

1886 – A frame hotel is built near the mouth of the cave. Managers of the hotel were Mr and Mrs W. Shelby. A livery stable is erected. The proprietor of the livery stable was W. Bryant Whitfield. This hotel, the Cave Hotel, is not the same as the Idaho Springs Hotel.

1887 – W.S. writes his initials and the date in the passage connecting the Junction Room with Great Relief Hall.

1887 – Lulu Bringhurst and Kate B. (Bringhurst ?) smoke their names and the date on the ceiling of Great Relief Hall.

1887 – A.G. Payne writes his name and the date on the wall in the Indian Trails Passage.

1887 – L. Whacey smokes his name and the date on the ceiling of Great Relief Hall.

1887 (March 9) – Borton & Roth smoke their names and the date on the ceiling of Great Relief Hall.

1887 (June 23) – There is a bran dance and picnic. According to the advertisement, there was "dancing day and night. The usual prices for admission and dancing will be charged. The mouth of the Cave will be brightly illuminated at night."

1887 (August) – J.M. Rice dies and A.B. Barbour of Bardstown, Kentucky buys the cave property for $19,815.

1887 (August 23) – Stock Show and Reunion. Special train was run each way between Cherry Station and Clarksville. "Will stop on Stacker's line within a few hundred yards of the cave."

1887 (November 3) – N. Gentry writes his name and the date on the wall near the Fallout Shelter supplies.

1888 – Walter Scott writes his name and the date on the wall of the Junction Room with black paint.

1888 – W. Scott writes his name and the date on the wall of the passage connecting the Junction Room with Great Relief Hall.

1888 – Richard Poston Jr writes his name and the date in Great Relief Hall.

1888 – Anna Goodsell writes her name and the date on the wall in Great Relief Hall.

1888 (May) – W.C. Stone, W.H. Patterson, W.H. Hill, H.T. Hill, and S.H. Neighbors carve their names and the date on the wall in Great Relief Hall.

1888 (June 29) – Baptist Picnic. "They came in two herdics, & furniture wagons, and a number of private conveyances. They picniced at noon and went into the cave."

1888 (July 11) – Geo Kirby, Kid Trimbly, Frannie Harrison, Susie Shelby, Liza Wilson, and Walley Jest ... (?) write their names and the date in black paint in the Foyer to the Counterfeiters Room.

Dunbar Cave

1888 (August 26) – Marye Tyler and Walter Scott write their names and the date on the wall in the Indian Trails Passage.

1888 (August 28) – Irwin Beaumot writes his name and the date on the wall of the Junction Room with black paint. The name Francis W. Smith appears to have been written at the same time.

1888 (December 13) – Jno. S. Low, C. Luther, and Miss L. Summer Lowe write their names and the date on the wall in the Indian Trails Passage.

1889 (May 14) – Miss Ella, Buky, and George Herdman, all of Rockwood, Tenn., leave their names and the date on the wall near Concession Rock.

1889 (May 14) – Miss Ella Buky, George Herdman, John Henery [*sic*] Tate write their names and the date in black paint in the foyer to the Counterfeiters Room.

1889 (June 4) – The Hotel is under the management of Mr and Mrs T.L. Yancey.

1890 – P.S. Barbour buys Dunbar Cave from O.K. Waller.

1890 (July 15) – The newspaper reports that: "The young people had an enjoyable time at the opening ball at Dunbar Cave Thursday night. Two bands furnished music."

1891 – John McPon writes his name and the date on the wall in the Indian Trails Passage.

1892 (July 8) – The Willapus Wallapus and the White Sheets meet at Dunbar Cave.

1892 (October 10) – The Dunbar Cave Hotel is destroyed by fire.

1893 – W.T. Haynes leases Dunbar Cave for the season. He give notice that a guide will be there during the summer.

1894 – P.T. Barbour dies. Ownership passes to his widow, Cecelia Barbour.

1902 (June 15) – George T. Fort writes his name and the date on the wall about 150 feet from the Entrance.

1903 (June 15) – Cecelia Barbour sells Dunbar Cave to the Electric Street Railway Company of Clarksville.

1903 – Laura Conrad writes her name on the wall of the Main Passage between the first room and the Counterfeiters Room.

1905 (August 15) – MH, EB, WN, and EW leave their initials and the date on the wall across from the steps leading up to the Counterfeiters Room.

1906 – Large amounts of fill dirt are placed in front of the entrance to create a level dance floor and the first wooden dance floor is built on top of this fill material.

1906 – R. Wells writes his name and the date on the wall of the Indian Trails Passage.

1906 (April 9) – Gill Patrick writes his name and the date on the wall of the Indian Trails Passage.

1907 (June 26) – Minnie Lou Sxoufldd writes her name and the date on the wall of the Indian Trails Passage.

1907 (July 4) – J.E. Lanchar writes his name and the date on the wall of the Indian Trails Passage.

1908 (August 18) – R.E. Lewis from Eunnis, Texas, writes his name and the date on the wall of the Indian Trails Passage.

1908 (August 18) – Kate Quarles from Hoptown, Kentucky, writes her name and the date on the wall of the Indian Trails Passage.

1910 (August 17) – W. Barnum, Wm. M. Case, and W.W. Case write their names and the date on the wall of the Indian Trails Passage.

1910 (August 24) – Ronald Murray writes his name and the date on the wall in the Indian Trails Passage.

1910 (August 24) – Austin Peay Jr and Ronald C. Murray write their names and the date on the wall in the Indian Trails Passage.

1910 (October 19) – Mr & Mrs R.F. Gill Jr write their names and the date on the wall in the Indian Trails Passage.

1911 (February 1) – W.W. Moore and L.C. Smith write their names and the date on the wall of the Indian Trails Passage.

1913 (October 15) – The Electric Street Railway Company of Clarksville sells their Dunbar Cave property to the Clarksville Dunbar Cave Railway Company.

1914 – Dunbar Cave is purchased by Wesley Drane and Austin Peay.

1914 – G.D. Hook writes his name and the date on the wall of the Indian Trails Passage.

1914 (July 9) – Christian Sunday School picnic at the cave.

1914 (August 31) – Louis Fisher sells his Dunbar Cave property to Wesley Drane and Austin Peay.

Dunbar Cave

1915 (December 21) – Wesley Drane and Austin Peay sell Dunbar Cave to J.H. Unself and H.B. Stout.

1915 (June 1) – The Idaho Springs Hotel opens it's new and up-to-date Summer Resort.

1915 (July 20) – Baptist Picnic at Dunbar Cave.

1915 (July 22) – Methodist Picnic at Dunbar Cave.

1915 (August ?) – Montgomery County Fair at Dunbar Cave.

1916 – Professor McQueen from Southwestern Presbyterian University (currently know as Austin Peay State University) builds a concrete dance floor in the mouth of Dunbar Cave. W.I. Haynes manages the cave while Mr and Mrs J.H. Tate run the Idaho Springs Hotel.

1916 (February) – Douglas Meriwether writes his name and the date on the wall of the Indian Trails Passage.

1916 (August 16) – E. Myers, M. Gaiser, and B. Meys write their names and the date on the wall of the Indian Trails Passage.

1917 (December 2) – Miss Mai Rose writes her name and the date on the wall near the entrance to the Indian Trails Passage.

1917 (December 2) – B. G. Hagewood and Emmett Farran write their names and the date on the wall near the entrance to the Indian Trails Passage.

1919 – C. D. Goggins writes his name and the date on the wall of the Indian Trails Passage.

1920 (June 12) – Season opens with a dance at the cave.

1920 (August 26-28) – County Fair held at Dunbar Cave.

1921 (June 11) – The National Quartet write their name and the date on the wall in the Indian Trails Passage.

1921 (July 7) – The Presbyterians have their annual picnic at Dunbar Cave.

1923 (June 20) – Opening dance for the season is held at Dunbar Cave. McCoy Miller and His Band from Lynchburg, Virginia perform. They stay and play each Tuesday and Friday night for the entire season.

1924 (July 4) – There is a big noon barbeque. Johnnie Ely and His Melody Boys furnished music for dancing.

1926 (July 4) – Independence Day Picnic sponsored by the American Legion has 4,000 to 5,000 paid admissions.

1929 – Mr & Mrs A.H. Aldred, Nashville, Tenn., write their names and the date on the wall of the Indian Trails Passage.

1929 (June 29) – Opening dance for the season is held at Dunbar Cave. Music is provided by Boddy Lewis' Golden Peacocks.

1931 – The Dunbar Cave and Idaho Springs Corporation is formed by John T. Cunningham, H.C. Merritt, Adolph Hach, J.T. Cunningham Jr, W.E. Beach, C.C. Beach, and B.F. Runyon.

1933 – The Idaho Springs Corporation adds electric lights to the cave, enlarges the lake, builds a concrete swimming pool, a bathhouse, tennis courts, a bowling alley, a concession stand, a bandstand, and extends the dance floor over the lake.

1933 – Francis Craig and His Orchestra, featuring Snooky Landman and Pee Wee, play at Dunbar Cave.

1933 (June 9) – Dinner at the Idaho Springs Hotel to be followed by a dance at the cave.

1933 (July 4) – The Fourth of July dance features Francis Carig and His Orchestra with Alpha Louis Morton.

1933 (July 15) – The Johnny Miller Orchestra plays at Dunbar Cave.

1934 (June 9) – King Oliver and his Recording Orchestra play at Dunbar Cave. "Yet's Dance, Chillun'"

1934 (June 11) – Huge Clarksville Sesqui-Centennial Celebration at Dunbar Cave with Governor Hill McAlister present. Francis Craig and His Band play that night.

1934 (June 16) – James Washington and the Hotel Ritz Orchestra play at Dunbar Cave.

1934 (July 17) – Jimmy Gallagher Orchestra plays at Dunbar Cave.

1934 (August 4) – Speaker of the United States House of Representatives Joseph W. Brynes speaks at Dunbar Cave. That night the Nations' Greatest Rhythm Orchestra, King Oliver and His Brunswick Recording Orchestra featuring Doris Duff, the Dilver Tenor, perform at the Post-Election Dance at Dunbar Cave.

1934 (August 14) – Beasey Smith's Orchestra of Nashville performs at Dunbar Cave.

1934 (August 28) – Jim Gallagher and His WSM Orchestra play at Dunbar Cave.

1934 (September 2) – Bernie Cummings Orchestra plays at Dunbar Cave.

1935 – D.L.S. writes his initials and the date on the wall in the Junction Room.

1935 – Earl Slate writes his name and the date on the wall in the Indian Trails Passage.

Dunbar Cave

1935 (January 15) – Jos. Slate writes his name and the date on the wall in the Indian Trails Passage.

1935 (May 31) – Dunbar Cave opens for the season with dinner and a dance. Dale Stevens and His Leviathan Orchestra perform.

1935 (June 15) – Johnny Hamp and His Orchestra perform.

1935 (July 4) – 5,000 people are present for the Fireworks Display and the Dance.

1935 (September 7) – Cabaret Dance.

1935 (September 10) – "Fat" Kelly and his Orchestra perform. Admission: Gentlemen 50 cents – Ladies 25 cents.

1936 – BJS, NGS, and EZS leave their initials and the date on the wall in the Indian Trails Passage.

1936 – Harry Eads scratches his name and the date on the wall of Great Relief Hall.

1936 – Sm. M. Paschall writes his name and the date on the wall of Great Relief Hall.

1936 – Irene Stato writes her name and the date on the ceiling of the Indian Trails Passage.

1936 (June 10) – Kay Kyser, the Dean of The Kollege of Musical Knowledge, and his band, featuring vocalists Ginnie Simms, Harry Babbit, and Ishcabibble play to a record crowd.

1936 (July14) – Johnny Hamp and his band play at Dunbar Cave.

1936 (July 22) – Ace Brigade and His Band perform at Dunbar Cave.

1936 (August 22) – Gus Arnheim and his Orchestra perform. Admission: Advance tickets $2.00 – At the Gate $2.50.

1936 (September 1) – The Arnheim Band plays, followed by Noble Sissle and His Orchestra with Lena Horne as the female vocalist.

1937 – During the off season, from 1936 to 1937, new concession stands, complete with a roof garden were built. The dance floor was enlarged, creating a lower level for private parties. The Dunbar Cave entrance now had it final, modern look.

1937 (June 17) – The Freddie Martin Band plays at Dunbar Cave.

1937 (June 29) – Dick Cistle and His Band perform at Dunbar Cave.

1937 (July 9) – Little Jack Little and His Orchestra play at Dunbar Cave.

1938 (May 30) – The Dunbar Cave season opens. Roy Acuff, Minnie Pearl, and Rod Brasfield perform.

1938 (June 10) – The Johnny Hamp Band plays at Dunbar Cave.

1938 (July 29) – During the day, there is a Grand Ole Opry Celebration with an Old Fiddlers Contest. That night the Henry King Orchestra played in the cave entrance.

1938 (August 2) – The Woody Herman Band plays at Dunbar Cave.

1938 (August 20) – Artie Shaw and his Band play at Dunbar Cave, with Billie Holiday singing.

1938 (August 26 & 27) – The Miss Tennessee Pagent is held at the Dunbar Cave Resort. Isham Jones and His Band performed both nights.

1939 (May 18) – Dunbar Cave opens for the season with W.E. Beach as manager.

1939 (June 14) – The opening dance features Francis Craig. Dinner at the Hotel is $1 per plate and dancing is $1 per person.

1939 (June 27) – Professional Square Dance Exhibition.

1939 (October 14) – Charles B. Staton, R.H. (Roy) Harris, and Billy Watts locate the Roy Woodard Entrance and explore the cave for several hundred feet.

1940 (January 26) – R.A. Burke writes his name and the date on the wall of the Indian Trails Passage.

1940 (June 11) – Ozzie Nelson and His Orchestra plays at Dunbar Cave. Harriet Hilliard was the vocalist. Ozzie and Harriet went on to have a hit TV show. The show was on ABC from October 3, 1952, to September 3, 1966, a fourteen season run.

1940 (July 1) – The Brass Band of the I.O.O.F. Home gives a concert at Dunbar Cave.

1940 (July 2) – George Olsen and his "Music of Tomorrow" perform on the new bandstand.

1940 (July 17) – Freddy Martin and his Orchestra perform. Admission: $1.25 in advance, $1.50 at the Gate.

1940 (August ?) – Ella Fitzgerald and Her Band perform at Dunbar Cave.

1940 (August 23 & 24) – The Tobacco Festival is held at Dunbar Cave. Art Kassel and His Orchestra played both evenings.

1940 (September ?) – The Del Courtney Orchestra plays the last concert of the season.

1941 (April 27) – B.H. Rice writes his name and the date on the wall of the Indian Trails Passage.

Dunbar Cave

1941 (May ?) – The Pinky Tomlin Band opens the season at Dunbar Cave.

1941 (June ?) – The Woody Herman Band plays at Dunbar Cave.

1941 (July ?) – The Jan Garber Band plays at Dunbar Cave.

1941 (August ?) – Isham Jones performs at Dunbar Cave.

1941 (August ?) – Blue Barron plays at Dunbar Cave.

1941 (August 19) – Lou Breeze performs at Dunbar Cave.

1941 (September 6) – Herbie Kaye plays at Dunbar Cave.

1942 – E. M. Flowers writes his name and the date on the wall near the entrance to the Indian Trails Passage.

1942 (May 15) – Dunbar Cave opens for the season. Season tickets for the Gate are $4.40.

1942 (June 12) – Francis Craig plays at Dunbar Cave. Admission is $1.65.

1942 (June 16) – The Blue Barron performs at Dunbar Cave.

1942 (July 4) – Charlie Nadgy's Band plays at Dunbar Cave.

1942 (July 8) – Jan Garber's Band plays at Dunbar Cave.

1942 (July ?) – Ina Ray Hutton and Her Playboys perform at Dunbar Cave.

1942 (August ?) – Will Osborne and His Orchestra play at Dunbar Cave.

1942 (August ?) – Clyde Lucas and His Band perform at Dunbar Cave.

1942 (September 9) – Jack Stalcup and His Band play at Dunbar Cave.

1943 (May 22) – Dunbar Cave and the Idaho Springs Hotel were closed until further notice.

1943 (July 6) – Dunbar Cave and Idaho Springs open for the season, but the swimming pool remains closed.

1943 (June 24) – The swimming pool opens for the season. The pool is now filled with spring water, not lake water as in previous years.

1944 (June 7) – Opening dance with Francis Craig and His Orchestra.

1944 (June 20) – Barney Rapp and his 13-piece Orchestra play at Dunbar Cave. "Due to curfew for the military, time of dance will begin at 8 o'clock and conclude at midnight."

1945 – William Kleeman purchases the Dunbar Cave Resort.

1945 (May 22) – Dunbar Cave opens for the season. Francis Craig plays for the Opening Dance.

1945 (August 17) – Francis Craig and his NBC Orchestra perform on this Friday night from 9:30 until 1:30.

1947 (July) – Nell R. Vaughan writes his name and the date on the wall at the entrance to the Indian Trails Passage.

1947 (July 17) – Aubrey M., from Clanton, Alabama, Bernie Roake, Bobby Roake, Dolores Roake, Faye Roake, and Ruth Providence write their names and the date on the wall at the entrance to the Indian Trails Passage.

1948 (April 26) – Roy Acuff purchases the Dunbar Cave Resort. He adds a golf course and an arena. Square dances are held on Tuesday and Friday nights and Round dances (Big Band music) are held on Saturday nights. Grand Ole Opry musicians perform at the cave on Sundays.

1948 (June 13) – Roy Acuff and His Smoky Mountain Boys give a special show. Mrs. Charlie C. Ledfore is Hostess for the Hotel.

1948 (July 1) – Roy Acuff announces that he will run for Governor on the Republican Ticket.

1948 (July 4) – Roy Acuff and His Smokey Moutain Boys perform, along with Slim Wooten and His Colorado Mountain Boys. Big fireworks show.

1948 (November 2) – Roy Acuff is defeated in his run for Governor of Tennessee.

1949 – Scott Acuff is manager of Dunbar Cave.

1950 (July 3) – Tony Pastor performs at Dunbar Cave.

1951 – Eddy Arnold and His Tennessee Plow Boys play at Dunbar Cave.

1954 (May 22) – Count Basie and his internationally famous Orchestra perform. Admission: $2.00 per person.

1955 (July 31) – Second Annual Ruffin Reddy Day. "Big Stage Show – 3:00 P.M. Eddie Hill – in person – Presents: Roy Acuff and his Smoky Mountain Boys. Ruffin Reddy – star of WSM TV." Visit Dunbar Cave: Adults $1.00, Children 12 and Under 50 cents.

1957 – Wayne Teeple writes his name, the date, and the notation "Cave Guide" on the wall in the Junction Room.

1958 (May 7) – Dunbar Cave opens for the season. "Gala Grand Ole Opry show at 3:00 P.M. will feature Roy Acuff and his Smoky Mountain Boys."

Dunbar Cave

1958 (May 20) – Woody Hermand and his Orchestra perform.

1959 (May 10) – Dunbar Cave opens for the season. "One big show of Grand Ole Opry will be held at 3:00 P.M."

1960 (August 14) – The Dunbar Cave Golf Course officially opens.

1961 – Broyce Hutchison and His Orchestra play at Dunbar Cave.

1961 (June 9-11) – Dunbar Cave Golf Club Invitational with Dizzy Dean.

1963 – McKay King purchases the Dunbar Cave Resort (but not the cave?).

1963 – Dan McDowell visits Dunbar Cave for the first time.

1963 (June 8) – The Apogees provide music for a dance at Dunbar Cave.

1965 (June 29) – Roy Acuff agrees to let Dunbar Cave be used as a Fallout Shelter. It will be the largest in Montgomery County and will accommodate 2,603 people for 2 weeks.

1965 (September 17) – 150 people spend Friday night in the Dunbar Cave Fallout Shelter as part of a Civil Defense exercise.

1966 (November 21) – McKay King buys Dunbar Cave under the name Swan Lake Development Corporation. King runs the cave and the pool and his wife Catherine runs the golf course.

1967 – The Dunbar Cave swimming pool is closed.

1971 – McKay King dies, leaving the Dunbar Cave Resort to his widow.

1971 – Dunbar Cave closes as a commercial cave.

1972 – City of Clarksville purchases the 300 acre Swan Lake Golf Course.

1973 (November 2) – Dunbar Cave is purchased by the state of Tennessee and becomes a State Natural Area.

1976 – Dan McDowell returns to Dunbar Cave.

1977 (March 7 – 9) – Tennessee Division of Archaeology excavates the entrance area of Dunbar Cave.

1977 (March 11) – The Northern Indiana Grotto begins the survey of Dunbar Cave.

1977 (May 14) – The Northern Indiana Grotto has its second surveying trip in Dunbar Cave.

1977 (June 25) – The Northern Indiana Grotto discovers 928 feet of virgin cave, which they name the Bat Bone Passage.

1978 (January 12) – The Northern Indiana Grotto completes the map of the Historic Section of Dunbar Cave. 16,145 feet of cave have been surveyed.

1978 (April 30) – Jack Countryman tries to push the upstream sump, with no luck.

1978 (May 14) – Clarence "Bud" Dillon and Steve Maegerlein use scuba gear to go upstream through the sump and discover a mile of virgin cave.

1978 (May 27) – Walter Scheffrahn, Jack Countryman, and Al Goodrich locate the Roy Woodard Entrance to Dunbar Cave.

1978 (June 15) – Tennessee Division of Archaeology begins another excavation of the entrance area of Dunbar Cave.

1978 (June 24) – Jack Countryman, Dan McDowell, and Roger Cole explore upstream in the Roy Woodard section of the cave.

1978 (July 3) – The Northern Indiana Grotto begins surveying the Roy Woodard portion of Dunbar Cave.

1978 (September 30) – The Tennessee Division of Archaeology's second entrance excavation is completed.

1978 (October 28-29) – The Northern Indiana Grotto surveys over 3,000 feet of passageways.

1978 (November 24 – 25) – The Northern Indiana Grotto discovers and surveys 2,100 feet of virgin cave.

1978 (December 31) – The Northern Indiana Grotto's survey stands at 8,100 feet.

1981 (January 3-5) – Dale J. Purchase, with the help of Bruce Herr, John Garner, and others, enters the Dunbar Cave Entrance, scuba dives through the sump and exits via the Woodard Entrance. This is the first Dunbar to Woodard trip ever made.

1985 (September 1) – The Northern Indiana Grotto completes its survey of Dunbar Cave. The total length is 42,594 feet or 8.066 miles.

1986 (March 28-30) – The Northern Indiana Grotto builds a gate on the Woodard Entrance.

1989 (May 20) – The Northern Indiana Grotto presents the final draft of the Dunbar Cave map to the state.

1993 (January 22) – Northern Indiana Grotto members discover the entrance to Double Dead Dog Drop.

Dunbar Cave

2003 (July 12) – Double Dead Dog Drop is connected to Dunbar Cave by explorers.

2005 (January 15) – Larry E. Matthews, Amy Wallace, Billyfrank Morrison, and Joe Douglas discover Indian Glyphs in Dunbar Cave.

2005 (May) – The National Speleological Society publishes the First Edition of *Dunbar Cave: The Showplace of the South*, Larry E. Matthews, Author and Thomas G. Rea, Editor.

2006 (July 29) – A ceremony is held at the entrance to Dunbar Cave to dedicate the new gate and to announce the discovery of ancient petroglyphs and pictographs in the cave.

2007 (October 9) – Larry E. Matthews begins research on the names and dates written on the walls and ceilings of Dunbar Cave. He is assisted by Amy Wallace and Bill Halliday. Bill Halliday discovers an etched drawing of a rattlesnake along with other etchings that he believes are prehistoric art.

2007 (December 12) – Larry E. Matthews visits Dunbar Cave to continue research on the names and dates written on the walls and ceilings. He is assisted by Amy Wallace and Bill Halliday.

2008 (November 26) – Larry E. Matthews visits Dunbar Cave to continue research on the names and dates written on the walls and ceilings. He is assisted by Amy Wallace, Billyfrank Morrison, and Adam Neblett.

2009 (September 16) – Larry E. Matthews visits Dunbar Cave to continue research on the names and dates written on the walls and ceilings. He is assisted by Amy Wallace and Adam Neblett.

2010 – The state of Tennessee closes Dunbar Cave due to the discovery of one solitary bat with White Nose Syndrome (WNS).

2011 (June 11) – Larry E. Matthews presents a program on the history of Dunbar Cave to the Montgomery County Historical Society. The meeting is held in the entrance to Dunbar Cave.

2011 – The National Speleological Society publishes the Second Edition of *Dunbar Cave: The Showplace of the South*, Larry E. Matthews, Author and Thomas G. Rea, Editor.

Appendix E

Suggested Additional Reading

If you enjoyed this book you may wish to read other books on caves and cave exploration. Here are a few suggestions for books that are still in print and readily available at the time that this book went to the printers:

Big Bone Cave, Larry E. Matthews, 2006, National Speleological Society, Huntsville, Alabama, 220 pages.

The fascinating story of America's largest saltpeter mine, located 50 miles north of Chattanooga. Mined in both the War of 1812 and the Civil War. Also the site of the discovery of giant ground sloth skeletons, an ice-age jaguar skeleton, and several American Indian mummies.

Blue Spring Cave, Larry F. Matthews and Bill Walter, 2010, National Speleological Society, Huntsville, Alabama, 346 pages.

The complete story of the exploration of Tennessee's longest cave and the 9th longest cave in the United States. In 1989 explorers enlarged a blowing crack at the end of what was only a 500-foot long cave and have since discovered 35 miles of virgin cave. Over a mile

of this cave is completely under water and was explored and mapped by cave divers.

Caves of Chattanooga, by Larry E. Matthews, 2007. National Speleological Society, Huntsville, Alabama, 198 pages.

The history of nine caves in the Chattanooga area that are either currently commercial caves, or that were commercial caves in the past.

Caves of Knoxville and the Great Smoky Mountains, by Larry E. Matthews, 2008. National Speleological Society, Huntsville, Alabama, 304 pages.

The history of fourteen caves in the Knoxville/Great Smoky Mountains area that are either currently commercial caves, or that were commercial caves in the past.

Caving Basics, Third Edition, Editor G. Thomas Rea, 1992. National Speleological Society, Huntsville, Alabama, 187 pages.

This book is an excellent introduction for new cavers who want to master the basic skills.

Cumberland Caverns, Third Edition, Larry E. Matthews, 2010, Greyhound Press, Cloverdale, Indiana, 196 pages. ISBN 978-0-9663547-5-1.

The complete history of the exploration of Tennessee's second longest cave, located 70 miles northwest of Chattanooga and open to the public.

On Call, Edited by John Hempel and Annette Fregeau-Conover, National Speleological Society, Huntsville, Alabama, 384 pages.

This is the most comprehensive book on cave rescue ever written.

On Rope II, by Allen Padgett and Bruce Smith, National Speleological Society, Huntsville, Alabama.

This is the most thorough guide available for rope techniques. These techniques are used for cave exploration, mountaineering, and rescue.

Speleology – Caves and the Cave Environment, George W. Moore and Nicholas Sullivan.

This is an easy-to-understand introduction to the science of caves, including their biology, geology, and how they form.

The Hidden World of Caves – A Children's Guide to the Underground Wilderness, Ronal C. Kerbo.

An excellent introduction to caves for younger readers.

There are several mail order firms that carry an extensive variety of books and magazines on caves. Most of these sources advertise in the *NSS News*. The National Speleological Society Bookstore carries a large number of caving books. A current list of what is available can be obtained by writing them at this address:

NSS Bookstore
2813 Cave Avenue
Huntsville, AL 35810-4431

Telephone: **256-852-1300**

You can also find the NSS Bookstore on the World Wide Web at:

http://nssbookstore.org

There are hundreds of books on caves that have been printed in the English language. You will have a wonderful selection to choose from at the above sites.

Index

Dunbar Cave

Wallace, Amy 3, 4, 6, 19, 21, 28, 33, 179
Ware, W. Porter 63
Warfield, Charles 38, 50
Warfield, C.P. 38, 43, 52, 53
War of 1812 41
Warriors Path 39
Warsaw Limestone 175
Watson, Patty Jo 17
Watts, Billy 136
Watts Trail 146, 147
Watts, Walter 136
Weathersby, Priscilla 33, 50, 53
Weekly Chronicle 42
Wells, Bob 3, 160
Western Indiana Grotto 147, 152
Wheel Barrow Track 51
Whidby, Jim 29, 39, 71
White Nose Syndrome 27, 173, 180, 192
White Place 148
White Sulphur water 68
Whitfield, W. Bryant 56

Willapus Wallapus 59, 65
Williams, Eleanor S. 61
Williams, Hank 85
Will Osborne and His Orchestra 80
Willoughby, Tom 166, 168, 169
Windy City Grotto 115
Winters, Ralph 53
Withers, David 3
Woodard, Roy 134
Woodbridge's Directory of Clarksville 53
Woodland Period 14, 26
Woodward, Cher 107
Woody Herman Band 79, 80
Wyandotte Cave, Indiana 39

Y

Yancey, T.L. 57
Young, Faron 81

Z

Zepp, Louise 4